Heat and Mass Transfer in Building Energy Performance Assessment

Heat and Mass Transfer in Building Energy Performance Assessment

Special Issue Editors

Robert Černý
Ákos Lakatos
Václav Kočí

MDPI • Basel • Beijing • Wuhan • Barcelona • Belgrade

MDPI

Special Issue Editors

Robert Černý
Czech Technical University
in Prague
Czech Republic

Ákos Lakatos
University of Debrecen
Hungary

Václav Kočí
Czech Technical University
in Prague
Czech Republic

Editorial Office
MDPI
St. Alban-Anlage 66
4052 Basel, Switzerland

This is a reprint of articles from the Special Issue published online in the open access journal *Energies* (ISSN 1996-1073) in 2019 (available at: https://www.mdpi.com/journal/energies/special_issues/ Building_Energy_Assessment).

For citation purposes, cite each article independently as indicated on the article page online and as indicated below:

LastName, A.A.; LastName, B.B.; LastName, C.C. Article Title. *Journal Name* **Year**, *Article Number, Page Range*.

ISBN 978-3-03921-926-1 (Pbk)
ISBN 978-3-03921-927-8 (PDF)

Contents

About the Special Issue Editors vii

Preface to "Heat and Mass Transfer in Building Energy Performance Assessment" ix

Jan Kočí, Václav Kočí and Robert Černý
A Method for Rapid Evaluation of Thermal Performance of Wall Assemblies Based on
Geographical Location
Reprinted from: *Energies* **2019**, *12*, 1353, doi:10.3390/en12071353 1

Imre Csáky, Tünde Kalmár and Ferend Kalmár
Operation Testing of an Advanced Personalized Ventilation System
Reprinted from: *Energies* **2019**, *12*, 1596, doi:10.3390/en12091596 17

**Panagiotis Stamatopoulos, Panagiotis Drosatos, Nikos Nikolopoulos and
Dimitrios Rakopoulos**
Determination of a Methodology to Derive Correlations Between Window Opening Mass Flow
Rate and Wind Conditions Based on CFD Results
Reprinted from: *Energies* **2019**, *12*, 1600, doi:10.3390/en12091600 30

Ákos Lakatos, Attila Csík, Anton Trník and István Budai
Effects of the Heat Treatment in the Properties of Fibrous Aerogel Thermal Insulation
Reprinted from: *Energies* **2019**, *12*, 2001, doi:10.3390/en12102001 51

Jan Fořt, Radimír Novotný, Anton Trník and RobertČerný
Preparation and Characterization of Novel Plaster with Improved Thermal Energy
Storage Performance
Reprinted from: *Energies* **2019**, *12*, 3318, doi:10.3390/en12173318 63

Astrid Tijskens, Hans Janssen and Staf Roels
Optimising Convolutional Neural Networks to Predict the Hygrothermal Performance of
Building Components
Reprinted from: *Energies* **2019**, *12*, 3966, doi:10.3390/en12203966 76

Lukáš Fiala, Michaela Petříková, Wei-Ting Lin, Luboš Podolka and Robert Černý
Self-Heating Ability of Geopolymers Enhanced by Carbon Black Admixtures at Different
Voltage Loads
Reprinted from: *Energies* **2019**, *12*, 4121, doi:10.3390/en12214121 94

About the Special Issue Editors

Robert Černý is a Full Professor of Physics and head of the Department of Materials Engineering and Chemistry at the Faculty of Civil Engineering, Czech Technical University in Prague, Czech Republic. He received his PhD and MEng degrees from Czech Technical University in Prague and his DSc degree from the Academy of Sciences of the Czech Republic. His research is devoted to building materials engineering and the theoretical and experimental analysis of heat, moisture, and chemical compounds transport in building materials. In particular, it includes physical and mathematical modeling of transport phenomena, solution of inverse problems, determination of heat-, moisture- and salt transport and storage parameters, and the investigation of material properties at high temperatures. He is the author or co-author of 438 publications indexed in the Web of Science database, with more than 3,000 citations and an h-index of 29. He also serves regularly as a chair or vice-chair of the international scientific committee of the Central European Symposium on Building Physics and Central European Symposium on Thermophysics.

Ákos Lakatos is an Associate Professor in the Faculty of Engineering at the University of Debrecen, where he is employed as the leader of the Building Physics Laboratory. He obtained his PhD in Physics in 2011. His main research fields are hygric measurements and thermal properties of different building and insulating materials and investigating the heat transfer in building structures to assess their energy efficiency. Nowadays, his research is deeply focused on super insulation materials. He was a Janos Bolyai research scholar of the Hungarian Academy of Sciences in 2015–2018 as well as a post-doctoral fellow of the Hungarian Government, Ministry of Human Capacities in 2013–2014. He was also a leader of a Consortium Partner from Hungary in a H2020 EU project Concerted Action Energy Performance of Buildings Directive (2014–2018). He has more than 50 publications indexed in the Web of Science database. His h-index is 12.

Václav Kočí is an Assistant Professor at the Department of Materials Engineering and Chemistry, Faculty of Civil Engineering, Czech Technical University in Prague. He obtained his PhD in Physical and Materials Engineering from Czech Technical University in Prague in 2013. His research activities are focused on numerical simulations of coupled heat and moisture transport in porous building materials, modelling of the biodegradation of building materials or corrections of errors by means of computational analysis of transport processes in experimental devices. With 68 publications indexed in the Web of Science database and 268 citations, his h-index is equal to 11

Preface to "Heat and Mass Transfer in Building Energy Performance Assessment"

The building industry is influenced by many factors and trends reflecting the current situation and developments in social, economic, technical, and scientific fields. One of the most important trends seeks to minimize the energy demand. This can be achieved by promoting the construction of buildings with better thermal insulating capabilities of their envelopes and better efficiency in heating, ventilation, and air conditioning systems.

Any credible assessment of building energy performance includes the identification and simulation of heat and mass transfer phenomena in both the building envelope and the interior of the building. As the interaction between design elements, climate change, user behavior, heating effectiveness, ventilation, air conditioning systems, and lighting is not straightforward, the assessment procedure can present a complex and challenging task. Simulations should then involve all factors affecting the energy performance of the building in question.

However, an appropriate physical model of heat and mass transfer for different building elements is not the only element that outputs of building energy simulations must consider. Boundary conditions in the form of weather data sets represent another crucial factor in determining uncertainties of the outputs. In light of current trends in climate change, this topic is vitally important.

Accordingly, this Special Issue aims to provide insights into recent advancements in experimental analyses, computational modeling, and in situ measurements with careful attention to assessments of building energy performance via identification of heat and mass processes in building enclosures and their assembly.

Kočí et al. [1] presented a method for the rapid evaluation of thermal performance of building envelopes without the need to use sophisticated and time-consuming computational modeling. The proposed approach was based on the prediction of monthly energy balances per unit area of a wall assembly using monthly averages of temperature and relative humidity, as well as the elevation of a building's location. Contrary to most other methods, the obtained results included moisture content effects on the thermal performance of walls. The developed formulas for calculation of monthly energy balances were verified for nine commonly used wall assemblies in Central Europe in ten randomly selected locations. The observed agreement of the predicated data was determined using advanced finite-element simulation tools and hourly climatic data, both of which make good prerequisites for the further application of method in both research and building practices.

Csáky et al. [2] investigated personalized ventilation systems in office buildings in the hope that they might bring important energy savings as well as an improvement of the indoor air quality and thermal comfort sensation of occupants at the same time. After analyzing eleven different air terminal devices, they presented in their paper the operation testing results of an advanced ventilation system. Based on the obtained air velocities and turbulence intensities, one of the devices was chosen to perform thermal comfort experiments with subjects. It was shown that, in the case of elevated indoor temperatures, thermal comfort can be improved considerably. A series of measurements were also carried out in order to determine the background noise level and the noise generated by the personalized ventilation system. It was shown that further developments of the air distribution system are needed.

Stamatopoulos et al. [3] presented a methodology for the development of an empirical equation which can provide the air mass flow rate imposed by single-sided wind-driven ventilation of a room, as a function of external wind speed and direction, using the results from Computational Fluid Dynamics (CFD) simulations. The proposed methodology is supposed to be useful for a wide spectrum of applications in which no access to experimental data or conduction of several CFD runs is possible, deriving a simple expression of natural ventilation rate, which can be further used for energy analysis of complicated building geometries in 0-D models or in object-oriented software codes. The developed computational model simulated a building belonging to Rheinisch-Westfälische Technische Hochschule (RWTH, Aachen University, Aachen, Germany) and its surrounding environment. A tilted window represented the opening that allowed fresh air to ventilate the adjacent room. The derived data from the CFD simulations for the air mass flow were fitted with a Gaussian function in order to achieve the development of an empirical equation. The numerical simulations were conducted using the Ansys Fluent v15.0® software package. In this work, the k-w Shear Stress Transport (SST) model was implemented for the simulation of turbulence, while the Boussinesq approximation was used for the simulation of the buoyancy forces. The coefficient of determination R^2 of the curve was in the range of 0.84–0.95, depending on the wind speed. This function provided the mass flow rate through the open window of the investigated building. Subsequently the ventilation rate of the adjacent room in air speed ranged from 2.5 m/s to 16 m/s without the necessity of further numerical simulations.

Lakatos et al. [4] focused their research on aerogels, which besides conventional insulations (plastic foams and wool materials), represent one of the most promising thermal insulation materials today. As one of the lightest solid materials available today, aerogels are manufactured through the combination of a polymer with a solvent, forming a gel. Fiber-reinforced types are mainly used for buildings. The changes in both the thermal performance and the material structure of the aerogel blanket were followed after thermal annealing. The samples were put under isothermal heat treatments at 70 °C for [HOW MANY?] weeks, as well as at higher temperatures (up to 210 °C) for one day. The changes in the sorption properties that resulted from the annealing were presented. Furthermore, the changes in the thermal conductivity were followed by a Holometrix Lambda heat flow meter. The changes in the structure and surface of the material due to the heat treatment were investigated by X-ray diffraction and with scanning electron microscopy. In addition, the above-mentioned measurement results of differential scanning calorimetry experiments were also presented. As a result of using equipment from different laboratories that support each other, it was found that the samples went through structural changes after undergoing thermal annealing. The aerogel granules separated down from the glass fibers and grew up. This phenomenon might be responsible for the change in the thermal conductivity of the samples.

Fořt et al. [5] tried to achieve building sustainability and energy efficiency via thermal energy storage systems based on latent heat utilization. They assumed that the application of phase change materials (PCMs) can substantially improve the thermal performance of building envelopes, decrease energy consumption, and support thermal comfort maintenance, especially during peak periods. On this account, the newly formed form-stable PCM (FSPCM) based on diatomite impregnated by dodecanol was used as an admixture for the design of interior plasters with enhanced thermal storage capability. In their study, the effect of FSPCM admixture on functional properties of plasters enriched by 8, 16 and 24 wt. % was determined. On this account, the physical, thermal, hygric, and mechanical properties were assessed in order to correlate obtained results with applied FSPCM dosages. Achieved results revealed only a minor influence of applied FSPCM admixture on material properties when compared to negative impacts of commercially produced PCMs. The differential scanning calorimetry disclosed variations of the phase change temperature, which ranged from 20.75 °C to 21.68 °C and the effective heat capacity increased up to 15.38 J/g accordingly to the applied FSPCM dosages.

Performing numerous simulations of a building component to assess, for example, its hygrothermal performance with consideration of multiple uncertain input parameters can easily become computationally inhibitive. To solve this issue, Tijskens et al. [6] replaced the hygrothermal model by a metamodel, a much simpler mathematical model which mimics the original model with a strongly reduced calculation time. In their paper, convolutional neural networks predicting the hygrothermal time series (e.g., temperature, relative humidity, moisture content) were used to that aim. A strategy was presented to optimise the networks' hyper-parameters, using the Grey-Wolf Optimiser algorithm. Based on this optimisation, some hyper-parameters were found to have a significant impact on the prediction performance, whereas others were less important. This approach was applied to the hygrothermal response of a massive masonry wall, for which the prediction performance and the training time were evaluated. The outcomes showed that with well-tuned hyper-parameter settings, convolutional neural networks were able to capture the complex patterns of the hygrothermal response accurately. Thus, they were well-suited to replace the time-consuming standard hygrothermal models.

Fiala et al. [7] aimed at sustainable development in the construction industry, which can be achieved by the design of multifunctional materials with good mechanical properties, durability, and reasonable environmental impacts. New functional properties, such as self-sensing, self-heating, or energy harvesting, are crucially dependent on electrical properties, which makes them inefficient common building materials. Therefore, various electrically conductive admixtures are used to enhance their electrical properties. Geopolymers based on waste or byproduct precursors are promising materials that can gain new functional properties by adding a reasonable amount of electrically conductive admixtures. The main aim of the Fiala et. al. paper lies in the design of multifunctional geopolymers with self-heating abilities. Designed geopolymer mortars based on blast-furnace slag activated by water glass and 6 dosages of carbon black (CB) admixture up to 2.25 wt. % were studied in terms of basic physical, mechanical, thermal, and electrical properties (DC). The self-heating ability of the designed mortars was experimentally determined at 40 and 100 V loads. The percolation threshold for self-heating was observed at 1.5 wt. % of carbon black with an increasing self-heating performance for higher CB dosages. The highest power of 26 W and the highest temperature increase of about 110 °C were observed for geopolymers with 2.25 wt. % of carbon black admixture at 100 V.

These papers offer a broad view of the relevant, diversified, and challenging problems arising in heat and mass transport in building energy performance assessment. We would like therefore thank to all the authors and reviewers for the care taken in preparing and assessing the papers, as well as the proficiency of the staff at MDPI.

References

[1] Kočí, J.; Kočí, V.; Černý, R. A Method for Rapid Evaluation of Thermal Performance of Wall Assemblies Based on Geographical Location. *Energies* **2019**, *12*, 1353.

[2] Csáky, I.; Kalmár, T.; Kalmár, F. Operation Testing of an Advanced Personalized Ventilation System. *Energies* **2019**, *12*, 1596.

[3] Stamatopoulos, P.; Drosatos, P.; Nikolopoulos, N.; Rakopoulos, D. Determination of a Methodology to Derive Correlations Between Window Opening Mass Flow Rate and Wind Conditions Based on CFD Results. *Energies* **2019**, *12*, 1600.

[4] Lakatos, Á.; Csík, A.; Trník, A.; Budai, I. Effects of the Heat Treatment in the Properties of Fibrous Aerogel Thermal Insulation. *Energies* **2019**, *12*, 2001.

[5] Fořt, J.; Novotný, R.; Trník, A.; Černý, R. Preparation and Characterization of Novel Plaster with Improved Thermal Energy Storage Performance. *Energies* **2019**, *12*, 3318.

[6] Tijskens, A.; Janssen, H.; Roels, S. Optimising Convolutional Neural Networks to Predict the Hygrothermal Performance of Building Components. *Energies* **2019**, *12*, 3966.

[7] Fiala, L.; Petříková, M.; Lin, W.T.; Podolka, L.; Černý, R. Self-Heating Ability of Geopolymers Enhanced by Carbon Black Admixtures at Different Voltage Loads. *Energies* **2019**, *12*, 4121.

Robert Černý, Ákos Lakatos, Václav Kočí
Special Issue Editors

energies

MDPI

Article

A Method for Rapid Evaluation of Thermal Performance of Wall Assemblies Based on Geographical Location

Jan Kočí *, Václav Kočí and Robert Černý

Department of Materials Engineering and Chemistry, Faculty of Civil Engineering, Czech Technical University in Prague, 166 29 Prague 6, Czech Republic; vaclav.koci@fsv.cvut.cz (V.K.); cernyr@fsv.cvut.cz (R.Č.)
* Correspondence: jan.koci@fsv.cvut.cz

Received: 11 March 2019; Accepted: 7 April 2019; Published: 9 April 2019

Abstract: In this study, we present a method for the rapid evaluation of thermal performance of building envelopes without the need of using sophisticated and time-consuming computational modeling. The proposed approach is based on the prediction of monthly energy balances per unit area of a wall assembly using monthly averages of temperature and relative humidity, as well as the elevation of a building's location. Contrary to most other methods, the obtained results include how moisture content in the wall effects its thermal performance. The developed formulas for calculation of monthly energy balances are verified for nine commonly used wall assemblies in Central Europe in 10 randomly selected locations. The observed agreement of the predicated data was determined using advanced finite-element simulation tools and hourly climatic data, which makes for good prerequisites for the further application of the method in both research and building practices.

Keywords: building envelope; thermal performance; energy balance; temperature; relative humidity; elevation

---

1. Introduction

A fundamental objective of building enclosures is to protect occupants from weather effects. Therefore, each part of a building envelope needs to meet certain thermal requirements in order to create a comfortable interior environment. A good thermal performance of building envelopes is very important in order to minimize overall energy consumption. Besides industry and transportation, residential households are one of the largest energy consumers in the European Union (EU). According to an EU report [1], 25.4% of total energy is consumed by residential houses and 70% of that amount is represented by heating energy [2]. This means that significant energy savings can be achieved by improving the effectiveness of heating systems or thermal insulating capabilities of building envelopes, which are required by national thermal standards.

Currently, when a building envelope is being designed or assessed, the U-value (thermal transmittance) is mostly used to indicate its insulating capabilities. It is a basic quantity describing the thermal insulating capability of building constructions; its required values are prescribed for each part of a building in any European country [3–5]. The U-value calculation is based on steady-state loading conditions and mostly comes from laboratory measurements [6–9]. However, there are several drawbacks to its laboratory methods. First, the accuracy is not always sufficient, as the effect of moisture on thermal conductivity of building materials is often neglected [10–12]. Although the national standards define design values of thermal conductivity and assume a certain level of operational moisture content, the thermal conductivity of building materials can significantly differ in the real conditions due to weather effects and presence of liquid moisture. Second, external thermal loads on building walls in real conditions are not steady. This may be due to changing outdoor conditions, such as temperature, relative humidity, wind speed, precipitation, and/or solar radiation.

Since the climate is comprehended as a local variable, it is apparent that geographical location affects significantly the thermal performance of individual wall assemblies. For that reason, many research studies have been aimed at the investigation of differences between laboratory and on-site thermal performance of building materials, components, or whole buildings. Byrne et al. [13] pointed out that predicted values of heat loss using standardized assumed material properties of the existing structure do not reflect the actual values achieved in situ. Marchio and Rabl [14] compared the predicted and observed performance of selected houses and apartments in France. Branco et al. [15] compared predicted and performed heat consumption of low energy family house in Switzerland. Roels et al. [16] provided an extensive comparison of various assessment methods for on-site characterization of the overall heat loss coefficient. Ficco et al. [17] conducted experimental measurement of in situ U-values and compared them against the estimated ones from design data and field analyses. The traditional empirical rules or standardized methods for U-value calculation were thus found not to work effectively, which was due to the high variability of the environmental and material properties or insufficient quality of input data. For a more proper assessment of thermal performance of building enclosures, more advanced techniques were supposed to be incorporated. Therefore, some new approaches for determination of thermal performance were suggested. For example, Robinson et al. [18] outlined a new transient, straightforward, and low-cost method for estimating the thermal properties of wall structures. Byrne at al. [19] designed a facility for testing the thermal properties of wall samples under both steady and transient conditions. Perilli et al. [20] performed a numerical analysis of thermophysical behavior of cork insulation based on in situ experimental data. Some other advanced techniques were applied, for example, for the analysis of the effect of wind velocity on quantification of heat losses through building envelope thermal bridges [21], the estimation of overall heat loss coefficient [22], convective heat transfer coefficient of exterior surface of building walls [23], or the prediction of residential heating demands [24].

In this paper, a method for rapid quantification of thermal performance of exterior wall systems is designed, which is intended to provide the designers and engineers with a fast and efficient tool for thermal design of residential buildings. The approach is based on the development of formulas for the calculation of monthly energy balances that only use monthly averages of temperature and relative humidity and the elevation of building's location as input parameters, but can achieve similar accuracy as advanced computational methods utilizing robust finite-element simulation tools and hourly climatic data. At the development of the calculation formulas, climatic data for 50 locations across the Czech Republic are used as a training set. The data for other 14 Czech locations are utilized as a testing set in the first step of the verification procedure. Another set of weather data for 10 randomly selected European locations are obtained from the Meteonorm software [25] and is used in the second verification step. The application of the method is presented for nine common wall systems, but it can be extended to any other type of building wall.

2. Methods

2.1. Climatic Data

For the investigation of thermal performance of the analyzed wall assemblies, 64 locations across the Czech Republic were selected and the weather data from those locations were collected. To ensure the widest range of weather data possible, the selection covered both lowlands and mountains across the country. All weather data were obtained from the Czech Hydrometeorological Institute, which is the official authority for meteorology, climatology, hydrology, and air quality protection in the Czech Republic. All data were applied in the form of the Test Reference Year (TRY) [26–28]. The data included hourly values of temperature, relative humidity, precipitation, wind direction, wind velocity, diffuse and direct short-wave radiation, sky long wave emission radiation, and long wave emission radiation. The list of involved locations together with their elevations is shown in Table 1.

Table 1. List of applied weather data.

Location (Elevation)	Location (Elevation)	Location (Elevation)
1 Bělotín (306 m)	23 Pec p. Sněžkou (816 m)	45 Kocelovice (519 m)
2 Bílá Třemošná (322 m)	24 Praha–Karlov (261 m)	46 Kuchařovice (334 m)
3 Brod nad Dyjí (177 m)	25 Přerov (210 m)	47 Liberec (398 m)
4 Čáslav (238 m)	26 Přimda (743 m)	48 Luka (510 m)
5 Červená (748 m)	27 Smolenice (345 m)	49 Lysá Hora (1322 m)
6 České Budějovice (394 m)	28 Stříbro (412 m)	50 Ostrava (253 m)
7 Doksany (158 m)	29 Šerák (1328 m)	51 Praha–Ruzyně (364 m)
8 Domažlice (458 m)	30 Svratouch (734 m)	52 Přibyslav (533 m)
9 Dukovany (400 m)	31 Tábor (459 m)	53 Ústí n. Labem (375 m)
10 Harrachov (675 m)	32 Temelín (500 m)	54 Horní Bečva (565 m)
11 Heřmanův Městec (275 m)	33 Tuhaň (160 m)	55 Úpice (413 m)
12 Holenice (432 m)	34 Tušimice (322 m)	56 Šumperk (328 m)
13 Holešov (222 m)	35 Ústí nad Orlicí (402 m)	57 Krušovice (379 m)
14 Cheb (483 m)	36 Val. Klobouky (160 m)	58 Mladá Boleslav (221 m)
15 Ivanovice na Hané (243 m)	37 Velké Meziříčí (452 m)	59 Filipova Huť (1110 m)
16 Jindřichův Hradec (524 m)	38 Vír (473 m)	60 Bečov n. Teplou (535 m)
17 Košetice (534 m)	39 Zbiroh (476 m)	61 Hustopeče (201 m)
18 Kostelní Myslová (569 m)	40 Železná Ruda (866 m)	62 Kestřany (381 m)
19 Měděnec (828 m)	41 Brno–Tuřany (241 m)	63 Slaný (307 m)
20 Most (240 m)	42 Hradec Králové (230 m)	64 Město Albrechtice (498 m)
21 Nepomuk (471 m)	43 Churáňov (866 m)	
22 Olomouc (215 m)	44 Karlovy Vary (603 m)	

2.2. Studied Wall Assemblies

The investigation of thermal performance was carried out for various types of both historical and contemporary building enclosures commonly used in Central Europe. Load-bearing materials included concrete (C), ceramic brick (CB), advanced hollow bricks (AHB), and sandstone (S). The contemporary building envelopes were provided with different types of thermal insulation layers based on expanded polystyrene (EPS) and mineral wool (MW), while the historical masonry did not have any thermal insulation. Exterior plasters were chosen with respect to the material composition of the envelopes, such as lime-cement plaster (LC), renovation plaster for historical masonry (RPHM), or lime-pozzolan plaster that was specially developed for the advanced hollow bricks (LPC). On the interior side of all structures, 10 mm thick lime-cement plaster was assumed. The list of studied building enclosures is shown in Table 2.

Table 2. List of studied building enclosures. LC: Lime-cement plaster; LPC: advanced hollow bricks; RPHM: renovation plaster for historical masonry.

Building Env.	Load-Bearing Material	Thermal Insulation (100 mm)	Plaster (10 mm)
1	Ceramic brick (450 mm)	N/A	LC plaster
2	Ceramic brick (450 mm)	Expanded polystyrene	LC plaster
3	Ceramic brick (450 mm)	Mineral wool	LC plaster
4	Concrete (300 mm)	Expanded polystyrene	LC plaster
5	Concrete (300 mm)	Mineral wool	LC plaster
6	Advanced hollow brick (500 mm)	N/A	LPC plaster
7	Advanced hollow brick (500 mm)	Expanded polystyrene	LPC plaster
8	Sandstone (800 mm)	N/A	N/A
9	Sandstone (800 mm)	N/A	RPHM

2.3. Computational Simulation

In order to simulate heat transfer through investigated wall assemblies, a 1-D simulation of heat and moisture transport through the building walls exposed to the exterior environment was conducted. Further, to increase the accuracy of thermal performance, the study included moisture transport to the heat transfer modelling with a straight intention, as moisture can significantly influence heat transport and storage parameters of building materials. In the formulation of the mathematical model of heat and moisture transport in multicomponent porous building material systems, a modified version of the Künzel's [29] mathematical model was used [30]. The modification of the original model was motivated by the effort of increasing the numerical stability, output accuracy, and reducing the overall time of computation. The heat and moisture mass balance equations can be expressed as:

$$\frac{dH}{dT}\frac{\partial T}{\partial t} = div(\lambda grad T) + L_v div(\delta_p grad p_v) \tag{1}$$

$$\left[\rho_w \frac{dw}{dp_v} + (n - w)\frac{M}{RT}\right]\frac{\partial p_v}{\partial t} = div\left[D_g grad p_v\right] \tag{2}$$

where H (J·m^{-3}) is the enthalpy density, T (K) the absolute temperature, λ (W·m^{-1}·K^{-1}) the thermal conductivity, L_v (J·kg^{-1}) latent heat of evaporation of water, δ_p (s) the water vapor diffusion permeability, p_v (Pa) the partial pressure of water vapor in the porous space, ρ_w (kg·m^{-3}) the density of water, w (m^3·m^{-3}) the moisture content by volume, n (-) the porosity of the porous body, and M (kg·mol^{-1}) the molar mass of water vapor, and R (J·K^{-1}·mol^{-1}) is the universal gas constant. D_g (s) is the global moisture transport function defined as:

$$D_g = B \cdot D_w \rho_w \frac{dw}{dp_v} + A \cdot \delta_p \tag{3}$$

where A and B in Equation (3) are the membership functions defining the transition between particular phases of water, which can be formulated as:

$$B = \begin{cases} 0 & \varphi \in \langle 0; 0.9) \\ 32\left[\left(\frac{1}{p_{v2}-p_{v1}}\right)(p_v - p_{v1})\right]^6 & \varphi \in \langle 0.9; 0.938) \\ 1 - 32\left[\left(\frac{1}{p_{v2}-p_{v1}}\right)(p_{v2} - p_v)\right]^6 & \varphi \in \langle 0.938; 0.976) \\ 1 & \varphi \in \langle 0.976; 1\rangle \end{cases} \tag{4}$$

$$A = 1 - B, \tag{5}$$

where the partial pressures of water vapor p_{v1} and p_{v2} (Pa) define the transition region. In this paper, the values of p_{v1} and p_{v2} correspond to the values of relative humidity of 90% ($\varphi = 0.9$) and 97.6% ($\varphi = 0.976$), respectively.

The computational model was implemented into an HMS simulation tool (Heat, Moisture and Salt transport), which is based on the general finite element package SIFEL (Simple Finite Elements) [31]. Both tools were developed at Faculty of Civil Engineering, Czech Technical University in Prague. HMS has been successfully used and validated in the recent past [32]. Each wall assembly was discretized from 29 to 41 nodes depending on its thickness and finite element method was applied. For the solving of partial differential equation, a non-linear, non-stationary solver with adaptive time controller was employed. A short survey of basic physical, thermal, and hygric properties used in the computational simulations is presented in Tables 3 and 4, where the following symbols are used: ρ_v is the bulk density, ρ_{mat} is the matrix density, ψ is the total open porosity, λ is the thermal conductivity, c is the specific heat capacity, $\mu_{dry\text{-}cup}$ is the water vapor diffusion resistance factor in dry-cup arrangement, and κ_{app} is the apparent moisture diffusivity. Those parameters were used in the computational model (1)–(5), either directly (such as thermal conductivity) or recalculated (water vapor diffusion resistance factor into water permeability, bulk density, and specific heat capacity into derivation of enthalpy density). Some parameters in Tables 3 and 4 have informative character only (matrix density and open porosity). More details on the particular parameters, such as their dependence on moisture content, can be found in the original references listed in Table 5.

Table 3. Basic physical, thermal, and hygric properties of load-bearing materials. C: concrete; CB: ceramic brick; AHB: advanced hollow bricks; S: sandstone.

Material Parameter	AHB	CB	C	S
ρ_v (kg·m^{-3})	1389	1831	2380	2191
ρ_{mat} (kg·m^{-3})	2830	2581	2715	2668
ψ (%)	50.9	27.9	12.3	17.9
λ (W·m^{-1}·K^{-1})	0.084	0.59	1.66	2.77
c (J·kg^{-1}·K^{-1})	1052	825	672	628
$\mu_{dry\text{-}cup}$ (–)	12.8	22.1	15.8	11.6

Table 4. Basic physical, thermal, and hygric properties of thermal insulating and coating materials. MW: mineral wool (MW) and EPS: expanded polystyrene.

Material Parameter	MW	EPS	LC	LPC	RPHM
ρ_v (kg·m^{-3})	70	16.5	1244	1713	1637
ρ_{mat} (kg·m^{-3})	2260	1020	2480	2658	2478
ψ (%)	96.9	98.4	49.8	35.6	33.9
λ (W·m^{-1}·K^{-1})	0.356	0.037	0.30	0.669	0.664
c (J·kg^{-1}·K^{-1})	810	1570	1054	831	922
μ (–)	2.62	58.00	7.52	27.26	23.6

Table 5. List of sources of input parameters for computational simulation.

Material	Reference
Advanced hollow brick	[33]
Ceramic brick	[34]
Concrete	[35]
Sandstone	[36]
Mineral wool	[37]
Expanded polystyrene	[11]
Lime-cement plaster	[38]
Lime-pozzolan plaster	[38]
Renovation plaster for historical masonry	[39]

The exterior environment was simulated using weather data from stations listed in Table 1, whereas the interior conditions were kept at 21 °C and 55% of relative humidity during the whole year. The initial conditions were same as interior boundary. Each simulation was run for 10 years in order to avoid the results being affected by initial conditions. The data from the last year of the simulation were used for further analysis.

The solar radiation and precipitation are important factors affecting the heat flux at interior wall surface. Therefore, it is very important to include those effects into the computational model in order to be able to predict monthly heat fluxes affected by both precipitation and solar radiation. The orientation of the wall plays a crucial role regarding the solar radiation and precipitation loads. As the proposed methods evaluate the average values of monthly heat fluxes from north and south orientation, a detailed analysis needs to be performed in order to investigate the model accuracy, as well as to analyze the effect of precipitation and solar radiation. The results of such an analysis are shown in the Discussion section.

2.4. Assessment of Thermal Performance

The quantification of thermal performance of the studied walls was done on the basis of calculation of time development of heat flux density $q(t)$ on the interior surface of the construction during a year. The heat flux densities were determined as:

$$q(t) = \lambda_{ip}(w, t)\frac{\Delta T_e(t)}{\Delta x_e} \tag{6}$$

where $\lambda_{ip}(w,t)$ $(W \cdot m^{-1} \cdot K^{-1})$ is the moisture-dependent thermal conductivity of the interior plaster, Δx_e (m) is the thickness of the element adjoining to the face side of the wall in the main direction of the heat flux, and ΔT_e (K) is the temperature difference between the opposite sides of the element adjoining to face side of the wall in the main direction of the heat flux. When the heat flux density as a function of time during a year is known, the monthly energy balances EB_1–EB_{12} of the wall assembly can be evaluated as a sum of heat flux densities during individual months. In the calculations of monthly balances, the positive values represent monthly heat gains (i.e., the necessity of cooling to keep interior temperature at prescribed level), while negative values represent monthly heat losses (i.e., the necessity of heating). As individual months of the year contain different number of days, the calculated values of energy balances were normalized to 30-day period allowing mutual comparison between the months. Monthly periods were chosen as a compromise between computational efficiency, model accuracy, and data availability.

2.5. Identification Procedure

The identification of the relation between thermal performance of studied walls and the weather data of certain locations was based on the optimization procedure. The objective of that procedure was to minimize the difference between predicted and simulated thermal performance (monthly energy balances) by identifying unknown correlation coefficients. From the weather data listed in Table 1, locations 1 to 50 were used as a training set, i.e., set of data on which the identification was carried out. The remaining data from locations 51 to 64 were used as a testing set, i.e., those data were excluded from the identification procedure and once the identification was finished, they were used for the verification of identified correlation coefficients. In order to assure the highest simplicity possible, the predicting formula for monthly energy balance of each studied wall assemblies was optimized in the linear form as:

$$EB_{i,pred} = c_0 + c_1 \cdot T_i + c_2 \cdot RH_i + c_3 \cdot E \tag{7}$$

where $i = 1, 2, \ldots, 12$ indicates the month, $EB_{i,pred}$ $(W \cdot h \cdot m^{-2} \cdot month^{-1})$ is the predicted monthly energy balance of wall assembly in particular location, T_i is the average monthly temperature in particular location, RH_i the average monthly relative humidity in particular location, E is the elevation, and c_0–c_3 are correlation coefficients unique for each building wall listed in Table 2. The average monthly values of temperature and relative humidity for all locations are presented as supplementary data in Tables S1 and S2. The data on elevation were provided in Table 1.

As the identification procedure was based on minimization of the difference d between simulated and predicted monthly energy balances over multiple locations, the objective was to find such combination of c_0–c_3 for each of studied walls that fulfills

$$d = \min\left(\sum_{n=1}^{50}\sum_{i=1}^{12}\left\|EB_{i,sim} - EB_{i,pred}(c_0, c_1, c_2, c_3)\right\|\right) \tag{8}$$

where n is the location number (see Table 1) and $EB_{i,sim}$ (W·h·m^{-2}·month^{-1}) is the simulated monthly energy balance using computational model (1)–(5).

3. Results

3.1. Identification Procedure

Prior to the identification procedure, computational simulations of nine building envelopes that were exposed to the effect of environment in 64 different locations were conducted. Moreover, each building envelope was investigated in two different orientations—north and south—in order to include the effect of solar radiation and wind direction in each location. In total, 1152 simulations were carried out, which the monthly heat flux densities as a function of time were calculated from. The final heat flux density for each building wall and location was calculated as an average of north and south orientation and the monthly energy balance of each wall under different location was then calculated from monthly sequences of heat flux densities. The obtained results were normalized to 30-day period and in this way 576 input values for the identification procedure were generated.

In the identification procedure, the correlation coefficients for each wall assembly listed in Table 2 were identified on the training set that consisted of locations 1–50 from Table 1. Then, the identified correlation coefficients were verified on the testing set given by locations 51–64 from the same table. The identified correlation coefficients from the training phase together with the coefficient of determination (R-square) between simulated and predicted data are presented in Table 6. In this table, BE# refers to the building envelope numbers as listed in Table 2. The visual comparison between simulated and predicted data in the training phase of the identification is shown in Figure 1, where the studied wall assemblies are grouped into four categories by the load-bearing materials. Similar grouping is used further in the manuscript.

Table 6. Identified correlation coefficients.

BE#	c_0	c_1	c_2	c_3	% Error	R^2
1	−14144.67	809.76	−16.21	0.2373	2.04	0.9985
2	−4333.37	215.39	3.61	0.0716	3.27	0.9962
3	−4232.08	212.14	3.62	−0.0126	3.87	0.9948
4	−4921.03	251.87	1.85	0.0830	1.70	0.9990
5	−5538.32	254.04	10.81	0.0131	2.85	0.9968
6	−2929.06	133.68	5.68	0.0334	4.57	0.9926
7	−2367.72	94.08	8.34	0.0040	6.43	0.9852
8	−31272.49	1782.25	−35.56	0.7538	1.63	0.9991
9	−30679.90	1747.26	−33.68	0.5970	1.69	0.9990

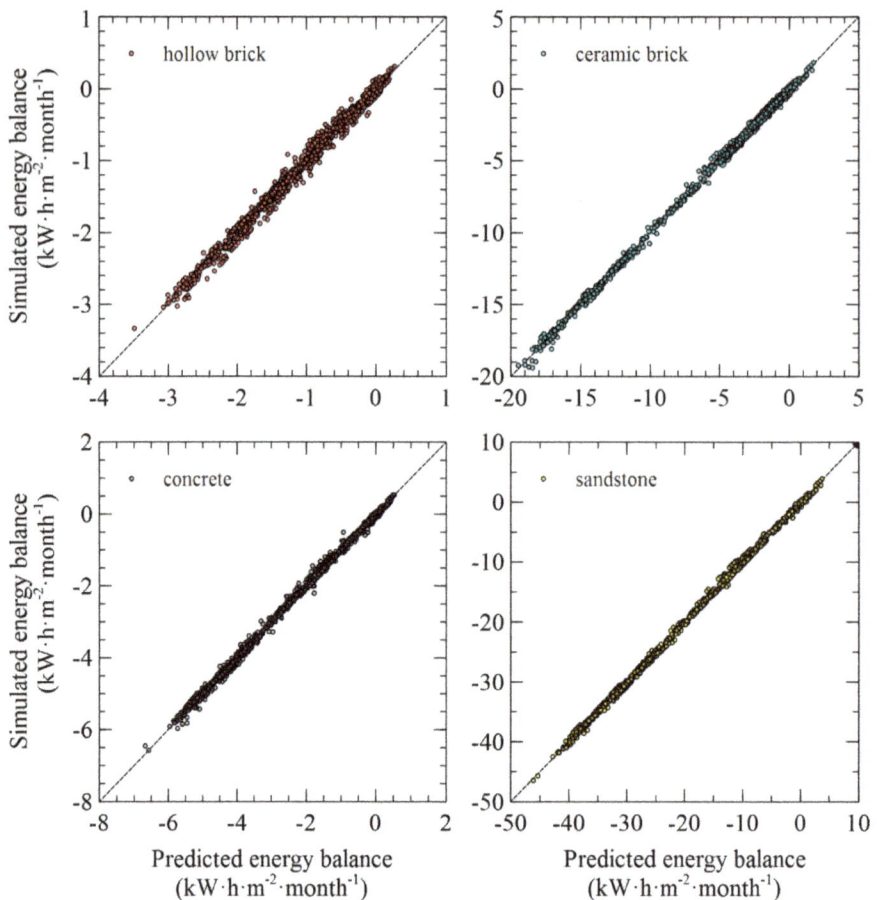

Figure 1. Training phase of the identification procedure.

It is obvious from Figure 1 that the identified correlation coefficients allowed to predict the monthly energy balance per unit area of the wall with a very high accuracy. Best results were achieved for sandstone masonry, while the worst agreement was observed for hollow brick masonry. The average prediction error between simulated and predicted values of monthly energy balance of individual wall assemblies was between 1.69% (sandstone masonry) and 6.43% (hollow brick masonry). The R-square was ranging between 0.9852 and 0.9991, which justified the utilization of linear formula in the identification procedure.

With the identified coefficients from Table 6, the testing procedure was run for each one of studied wall assemblies in order to verify the accuracy on blind data. This means, that remaining 14 localities, which were excluded from the training procedure were now tested with identified correlation coefficients c_0–c_3. The results of testing procedure are shown in Figure 2. The agreement between predicted and simulated data was very good, showing excellent setup of correlation coefficients. The average prediction error between simulated and predicted values of monthly energy balance of individual wall assemblies was between 1.46% (sandstone masonry) and 6.18% (hollow brick masonry). The R-square was ranging between 0.9860 and 0.9992.

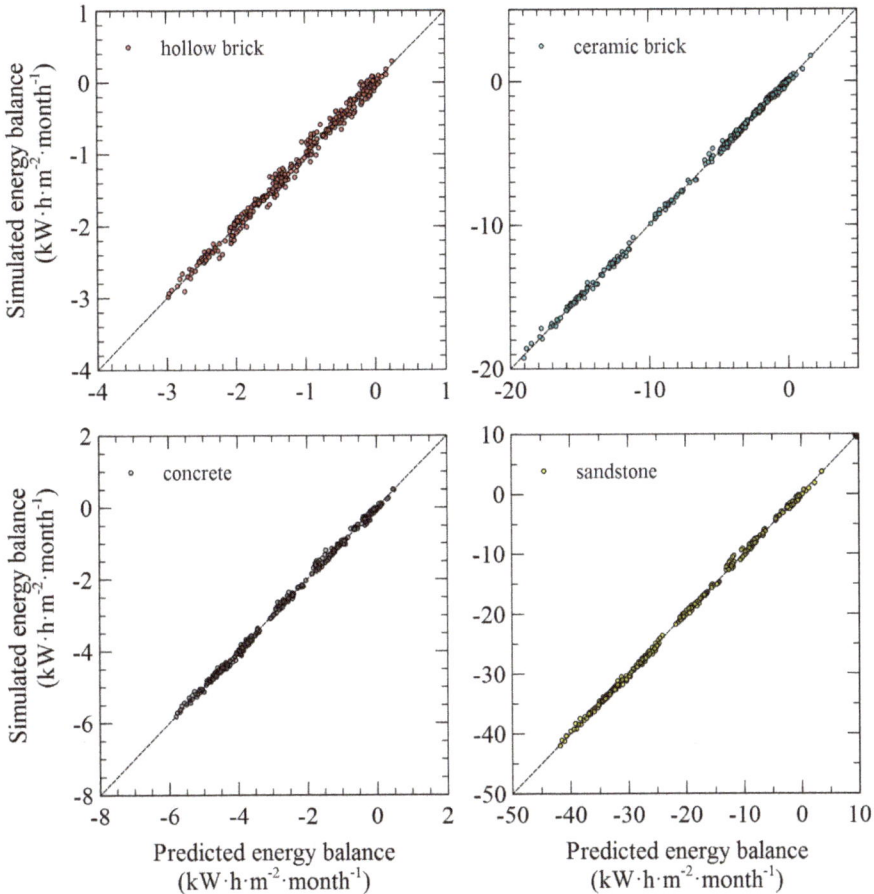

Figure 2. Testing phase of the identification procedure.

3.2. Verification Using Meteonorm Data

In order to support the results presented in the previous section, an additional verification procedure was carried out. This procedure was based on the application of the derived formulas for independent weather data obtained from the Meteonorm database [27] and comparison of predicted results with simulated data. For that purpose, 10 random locations from across Europe were selected, namely 1: Dublin (elevation 82 m), 2: Goteborg (20 m), 3: Helsinki (53 m), 4: Nantes (26 m), 5: Mannheim (106 m), 6: Warszawa (130 m), 7: Graz (342 m), 8: Nancy (212 m), 9: København (28 m), and 10: Štrbské Pleso (1368 m); from the weather data of these locations the average monthly values of temperature and relative humidity were exported. The input values, which are presented as supplementary data in Tables S3 and S4, were substituted into Equation (7), recalculated, and normalized into monthly energy balances and compared with simulated energy balances.

The results of the verification are shown in Figure 3. The agreement between predicted and simulated data was very good again, showing an excellent setup of correlation coefficients. The average prediction error between simulated and predicted values of monthly energy balance of individual wall assemblies ranged between 1.86% (concrete masonry) and 6.81% (hollow brick masonry). The R-square was ranging between 0.9834 and 0.9985.

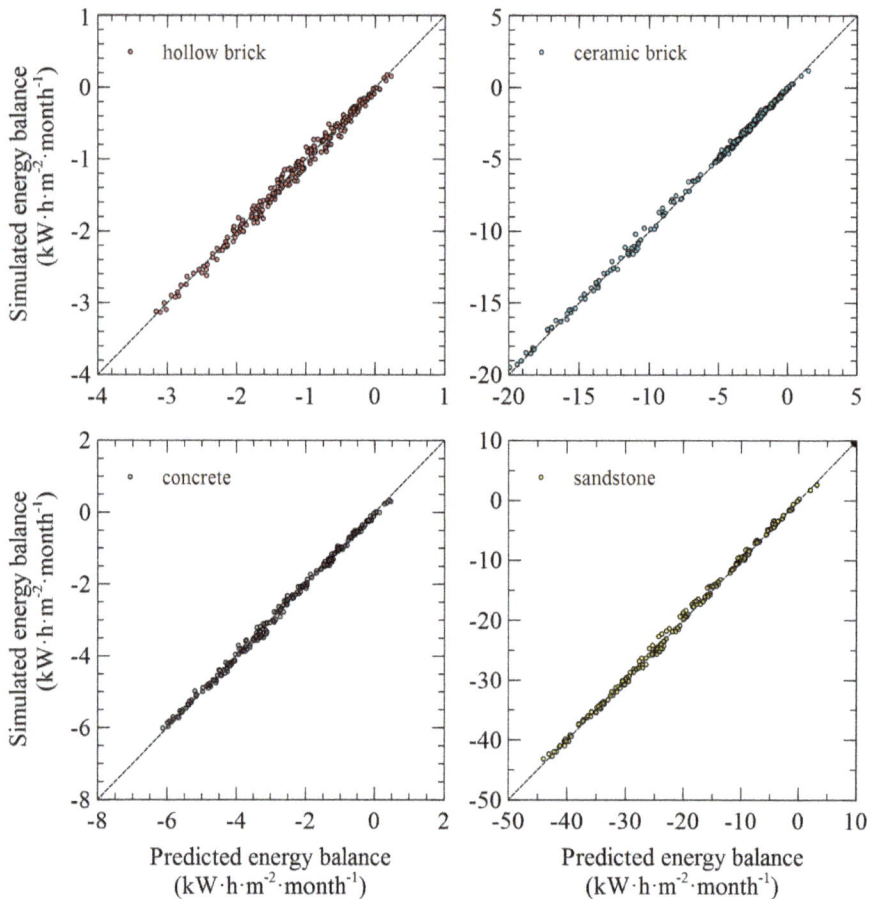

Figure 3. Verification using Meteonorm data.

4. Discussion

The results presented in Figures 1–3 show a good agreement between the data simulated using an advanced hygrothermal model and predicted by the proposed approach based on the knowledge of monthly temperature, relative humidity, and elevation. This means that in a real application the knowledge of commonly available weather statistics for a given location, together with a unique combination of c_0–c_3 coefficients from Equation (7) allows to produce a set of monthly energy balances capable of assessing the thermal performance of a wall assembly with a sufficient accuracy. Although for each wall assembly it takes 80 h of computational time to identify and verify obtained correlation coefficients, this method can be quite effective as the outputs can be used in simple algebra to obtain results that are comparable with those from sophisticated computational models. Moreover, the presented method can be extended to any kind of wall assembly or any other part of building envelope, such as glazing or roofs, providing the users with a tool that can produce results with research-like quality. However, it is important to say that finding those coefficients might be time-consuming and requires expert or research skills. The obtained U-values or heat fluxes can be used for fast assessment in such cases, where 1-D analysis is needed or requested. Moreover, the effective U-values (changing with time) can be used in advanced BIM models instead of standard U-values obtained from theoretical calculations (see below).

The presented analysis was done for continuous wall assemblies only as they are most typical structures in the Central Europe. However, the method can be extended to cavity/frame assemblies as well. In case that ventilated cavity or air gap need to be modelled, it will be necessary to replace the Künzel's mathematical model or to couple it with some other CFD model. Basically, this method should be comprehended as tailor-made, which is primarily dedicated to the very same wall assemblies as presented in this paper or for their very slight modifications. If an application on different kind of building envelope is needed, it is recommended rather to perform the simulation and optimization procedure from scratch than to approximate the results from available outputs. On the other hand, the presented method can be used to create input parameters for some approximation models, that will produce final and accurate outputs for various types of wall assemblies.

From the point of view of selected time-frame, the monthly averages seem to be most suitable choice for several reasons. First, the weather data are usually available for free without any additional costs, which is a good precondition for application of this method in practice. Second, considering the fact that national standards define only one value of thermal transmittance that is not changing over time, choosing monthly values gives a reasonable resolution for classifying building performance during the year. Additionally, lower time-frame would bring high fluctuations to the obtained results. Although the accuracy will be higher, it will be not suitable for practical applications.

The monthly values of energy balance may be effectively used for design of buildings' heating and cooling components or as advanced input parameters in more complex models used, e.g., for the assessment of energy efficiency of buildings or overall U-value.

Since the U-value is defined as the heat flux density through a given structure divided by the difference in environmental temperatures on either side of the structure in steady state conditions, the monthly energy balances can be simply used for calculation of equivalent or apparent U-values. When average monthly temperatures are known, each month can be considered as a steady-state period. Then, an apparent U-value can be calculated from monthly energy balance and used as a more accurate parameter describing the insulation capabilities of building elements. The apparent U-value can be calculated as:

$$U_{app} = \frac{1}{12} \sum_{i=1}^{12} \frac{1}{720} \frac{EB_i}{T_{i,int} - T_{ext,i}} \qquad (9)$$

where U_{app} (W·m^{-2}·K^{-1}) is the apparent U-value, EB_i (W·h·m^{-2}·month^{-1}) is the monthly energy balance calculated from Equation (7), $T_{i,int}$ (K) is the interior temperature (294.15 K), and $T_{i,ext}$ (K) is the average monthly exterior temperature. Since all the computational simulations in this research were conducted using an advanced hygrothermal model, the calculated outputs include the effect of moisture content on heat transport through the materials involved. This provides a higher accuracy than some common laboratory experiments or calculations done by more simplified techniques.

As an example of utilization of the proposed approach, a comparison between standardized and apparent U-value is provided below. In this example the wall assembly made of ceramic brick and polystyrene is chosen (building envelope 2, see Table 2), which is subjected to the effect of two environmental loads: Prague, Karlov, and Šerák (locations 24 and 29, see Table 1). The standardized procedure defines U-value as:

$$U = \frac{1}{R_{si} + R + R_{se}} \qquad (10)$$

where R (m^2·K·W^{-1}) is thermal resistance of the construction, R_{si} and R_{se} (m^2·K·W^{-1}) are external surface and internal surface resistances defined by the standards (according to [6], R_{si} = 0.13 m^2·K·W^{-1} and R_{se} = 0.04 m^2·K·W^{-1}). The R-value is calculated as:

$$R = \sum \frac{d_i}{\lambda_i} \qquad (11)$$

where d_i (m) is the thickness of individual layer in the composition of wall assembly and λ_i (W·m^{-1}·K^{-1}) is the thermal conductivity of the material involved in that layer. The U-value for building envelope 1 calculated according to (10) is equal to 0.258 W·m^{-2}·K^{-1}. The apparent value calculated from (9) using (7) and data from Table 6, Tables S1 and S2 is equal to 0.205 W·m^{-2}·K^{-1} for Prague and 0.246 W·m^{-2}·K^{-1} for Šerák. Although the standard U-value is on the safe side in this case, as it claims higher (i.e., worse) U-value than the apparent U-value approach, when individual months are analyzed in detail, it can be different in some other cases. Looking at Figure 4 showing apparent U-values during individual months, it is obvious that the construction will not meet the criteria given by standards during winter periods. The brick wall located in Prague will not stand the comparison in months December to March, while the same wall located in Šerák will not meet the criteria from November to March.

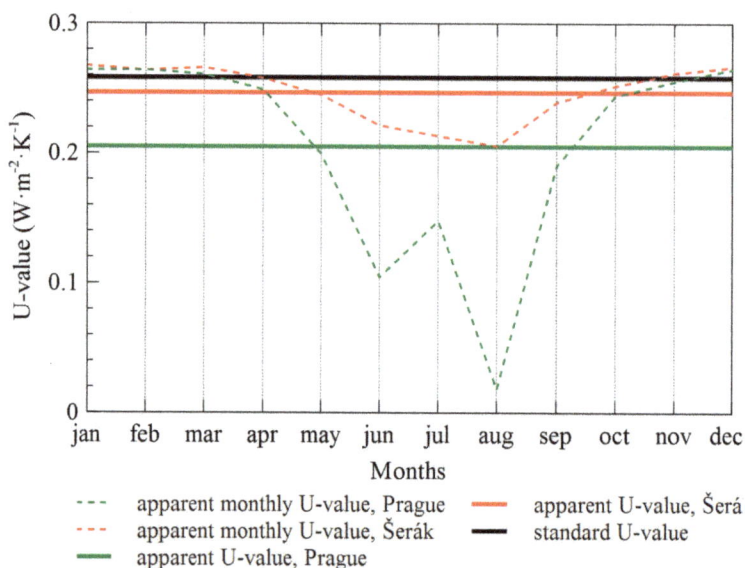

Figure 4. Comparison of standard and apparent U-values for brick masonry.

The analysis of solar radiation and precipitation and wall orientation was performed on wall assemblies made of ceramic brick both insulated and non-insulated (building envelopes 1 and 2). For that analysis, a location of Velké Meziříčí (location 37) was selected. The monthly heat fluxes for different wall orientations are depicted in Figure 5.

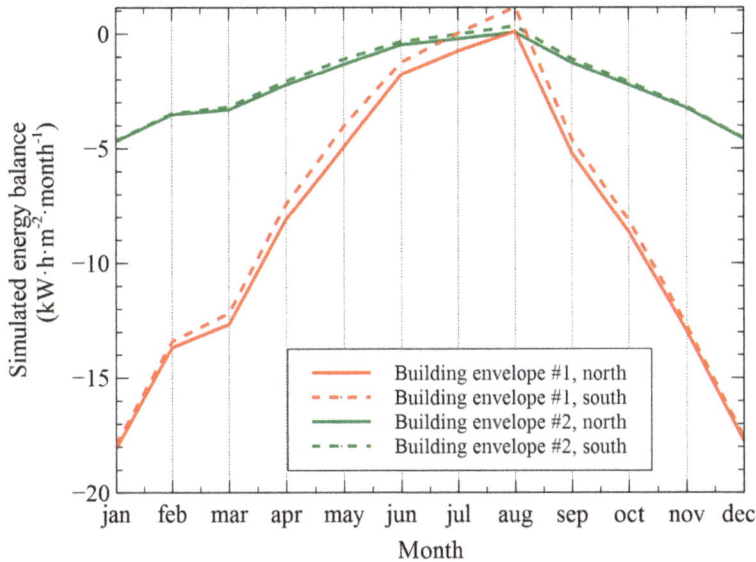

Figure 5. Comparison of monthly heat fluxes on interior surface of brick wall with different orientations.

The highest differences in simulated energy balances can be observed during summer period especially when non-insulated walls (building envelope #2) are considered. The differences in non-insulated walls (building envelope #1) range from 0.55% in winter (December) to more than 100% in summer (July). In absolute numbers, the differences are up to 1.050 kW·h·m^{-2}·month^{-1} (August). The differences in case of insulated brick wall are significantly less when speaking of absolute numbers. The highest difference in heat fluxes of north and south orientation can be observed in the same month of August, but the difference is only 0.269 kW·h·m^{-2}·month^{-1}, which is given by the insulation capability of expanded polystyrene. Such inaccuracies should be considered when using this method in the practice.

The effect of solar radiation (SR) and precipitation (PP) on monthly heat fluxes is shown in Figure 6. Similarly, to results shown in Figure 4, the highest differences can be observed in case of a non-insulated wall. The effects of solar radiation and precipitation contribute to the overall energy balance by approximately 2% in winter periods, but more significantly in summer. In case of non-insulated wall, the sun radiation can change the overall heat balance from negative to positive, which may be a very significant factor. For that reason, the effects of solar radiation and precipitation should be included in the computational model in order to produce satisfactory results. The fact that the presented method allows for the prediction of thermal performance of wall assembly including the effects of solar radiation and precipitation only from the knowledge of average monthly values temperature and relative humidity together with the elevation makes the method very useful. Since there can be found some level of correlation between temperature and solar radiation, or relative humidity and precipitation, the knowledge of limited weather data could be sufficient to bring relatively accurate predictions.

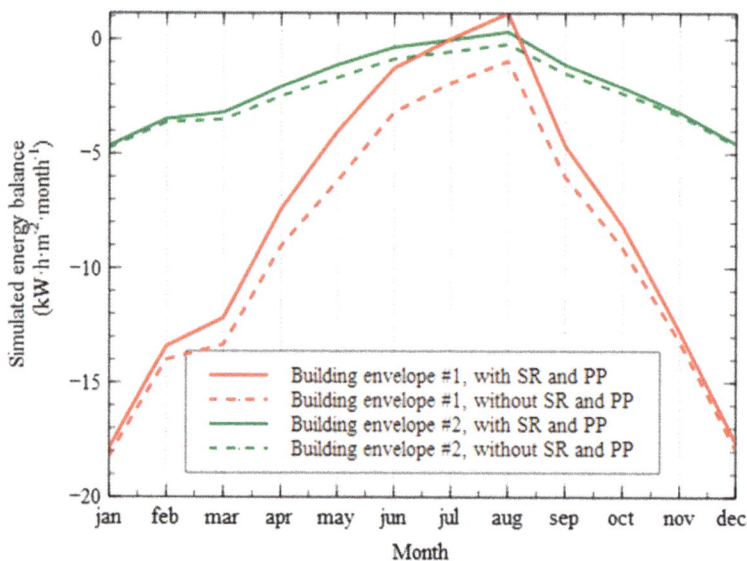

Figure 6. Analysis of the effect of solar radiation (SR) and precipitation (PP) on the simulated energy balance of south oriented walls.

5. Conclusions

In this study, we introduced a method suitable for rapid evaluation of thermal performance of building walls, which needs only monthly averages of temperature and relative humidity for a given location and its elevation as input data. The proposed approach was successfully tested for nine different types of wall assemblies. The results showed a good accuracy of the method; the average prediction error for tested wall assemblies was ranging between 1.63 and 6.43%.

The proposed approach can be considered as very time-saving, as compared with the methods that involve utilization of robust models. On the other hand, since the effect of moisture content is included in the model outputs, this approach outperforms more simplified models and methods. As a result, it offers a solution, which is neither too simple nor too complex. The presented method can be used for any location across Europe and can also be easily extended to any kind of wall assembly or building envelope component. Since the utilization of the proposed method is demonstrated on nine different wall assemblies only, the method should be extended to a broader range of wall assemblies or building components. In this way, a catalogue or database for civil engineers and designers can be generated, facilitating the thermal design of building structures or fast pre-assessment of wall assemblies from several points of view, e.g. predispositions to frost-induced damage, biofilms growth conditions or salt attack.

Supplementary Materials: The following are available online at http://www.mdpi.com/1996-1073/12/7/1353/s1, Table S1: Average monthly temperature of investigated locations. Table S2: Average monthly relative humidity of investigated locations. Table S3: Average monthly temperatures obtained from Meteonorm database. Table S4: Average monthly relative humidity obtained from Meteonorm database.

Author Contributions: Methodology, J.K.; Resources, J.K.; Software, V.K.; Supervision, R.Č.; Validation: V.K.; Writing–original draft, J.K.; Writing–review & editing, J.K.

Funding: This research was funded by the Czech Science Foundation, grant number 19-01558S.

Conflicts of Interest: The authors declare no conflict of interest.

Energies **2019**, *12*, 1353

References

1. Consumption of Energy. Available online: http://ec.europa.eu/eurostat/statistics-explained/index.php/Consumption_of_energy (accessed on 17 January 2018).
2. Cansino, J.M.; Pablo-Romero, M.d.P.; Román, R.; Yniguez, R. Promoting renewable energy sources for heating and cooling in EU-27 countries. *Energy Policy* **2011**, *39*, 3803–3812. [CrossRef]
3. ČSN 73 0540, *Thermal Protection of Buildings—Part 3: Design Value Quantities*; Czech Office for Standards, Metrology and Testing: Prague, Czech Republic, 2005.
4. UNE-EN 673 *Glass in Building—Determination of Thermal Transmittance (U Value)—Calculation Method*; Deutsches Institut für Normung: Berlin, Germany, 2011.
5. BS EN ISO 10077-2, *Thermal Performance of Windows, Doors and Shutters. Calculation of Thermal Transmittance; Numerical Method for Frames*; British Standards Institution: London, UK, 2017.
6. Caruana, C.; Yousif, C.; Bacher, P.; Buhagiar, S.; Grima, C. Determination of thermal characteristics of standard and improved hollow concrete blocks using different measurement techniques. *J. Build. Eng.* **2017**, *13*, 336–346. [CrossRef]
7. EN ISO 8990:1994, *Thermal Insulation—Determination of Steady State Thermal Transmission Properties—Calibrated and Guarded Hot Box*; International Organization for Standardization: Geneva, Switzerland, 1994.
8. ASTM C1363-11, *Standard Test Method for Thermal Performance of Building Materials and Envelope Assemblies by Means of a Hot Box Apparatus*; ASTM International: West Conshohocken, PA, USA, 2011.
9. ASHRAE Transactions 92, *Thermal Resistance Measurements of Well-Insulated and Super Insulated Residential Walls Using a Calibrated Hotbox*; ASHRAE Inc.: Atlanta, GA, USA, 1986; pp. 604–619.
10. Szodrai, F.; Lakatos, A.; Kalmar, F. Analysis of the change of the specific heat loss coefficient of buildings resulted by the variation of the geometry and the moisture load. *Energy* **2016**, *115*, 820–829. [CrossRef]
11. Jerman, M.; Černý, R. Effect of moisture content on heat and moisture transport and storage properties of thermal insulation materials. *Energy Build.* **2012**, *53*, 39–46. [CrossRef]
12. Khan, M.I. Factors affecting the thermal properties of concrete and applicability of its prediction models. *Build. Environ.* **2002**, *37*, 607–614. [CrossRef]
13. Byrne, A.; Byrne, G.; Davies, A.; Robinson, A.J. Transient and quasi-steady thermal behaviour of a building envelope due to retrofitted cavity wall and ceiling insulation. *Energy Build.* **2013**, *61*, 356–365. [CrossRef]
14. Marchio, D.; Rabl, A. Energy-efficient gas-heated housing in France: Predicted and observed performance. *Energy Build.* **1991**, *17*, 131–139. [CrossRef]
15. Branco, G.; Lachal, B.; Callinelli, P.; Weber, W. Predicted versus observed heat consumption of a low energy multifamily complex in Switzerland based on long-term experimental data. *Energy Build.* **2004**, *36*, 543–555. [CrossRef]
16. Roels, S.; Bacher, P.; Bauwens, G.; Castano, S.; Jiménez, M.J.; Madsen, H. On site characterisation of the overall heat loss coefficient: Comparison of different assessment methods by a blind validation exercise on a round robin test box. *Energy Build.* **2017**, *153*, 179–189. [CrossRef]
17. Ficco, G.; Iannetta, F.; Ianniello, E.; Romana d'Ambrosio Alfano, F.; Dell'sola, M. U-value in situ measurement for energy diagnosis of existing buildings. *Energy Build.* **2015**, *104*, 108–121. [CrossRef]
18. Robinson, A.J.; Lesage, F.J.; Reilly, A.; McGranaghan, G.; Byrne, G.; O'Hegarty, R.; Kinnane, O. A new transient method for determining thermal properties of wall section. *Energy Build.* **2017**, *142*, 139–146. [CrossRef]
19. Byrne, A.; Byrne, G.; Robinson, A. Compact facility for testing steady and transient thermal performance of building walls. *Energy Build.* **2017**, *152*, 602–614. [CrossRef]
20. Perilli, S.; Sfarra, S.; Guerrini, M.; Bisegna, F.; Ambrosini, D. The thermophysical behaviour of cork supports doped with an innovative thermal insulation and protective coating: A numerical analysis based on in situ experimental data. *Energy Build.* **2018**, *159*, 508–528. [CrossRef]
21. O'Grady, M.; Lechowska, A.A.; Harte, A.M. Quantification of heat losses through building envelope thermal bridges influenced by wind velocity using the outdoor infrared thermography technique. *Appl. Energy* **2017**, *208*, 1038–1052. [CrossRef]
22. Olofsson, T.; Andersson, S. Overall heat loss coefficient and domestic energy gain factor for single-family buildings. *Build. Environ.* **2002**, *37*, 1019–1026. [CrossRef]

23. Liu, Y.; Harris, D.J. Full-scale measurements of convective coefficient on external surface of a low-rise building in sheltered conditions. *Energy Build.* **2007**, *42*, 2718–2736. [CrossRef]

24. Wang, Z.Y.; Srinivasan, R.S.; Shi, J. Artificial Intelligent Models for Improved Prediction of Residential Space Heating. *J. Energy Eng.* **2016**, *142*, 04016006. [CrossRef]

25. *Meteonorm, Version 6.0, Software Version 6.1.0.20 from April 2010;* Meteotest: Bern, Switzerland, 2010.

26. Bilbao, J.; Miguel, A.; Franco, J.A.; Ayuso, A. Test Reference Year Generation and Evaluation Methods in the Continental Mediterranean Area. *J. Appl. Meteorol.* **2004**, *43*, 390–400. [CrossRef]

27. Kalamees, T.; Kurnitski, J. Estonian test reference year for energy calculations. *Proc. Estonian Acad. Sci. Eng.* **2006**, *12*, 40–58.

28. Lee, K.; Yoo, H.; Levermore, G.J. Generation of typical weather data using the ISO Test Reference Year (TRY) method for major cities of South Korea. *Build. Environ.* **2010**, *45*, 956–963. [CrossRef]

29. Künzel, H.M. Simultaneous Heat and Moisture Transport in Building Components. Ph.D. Thesis, IRB Verlag, Stuttgart, Germany, 1995.

30. Maděra, J.; Kočí, J.; Kočí, V.; Kruis, J. Parallel modelling of hygrothermal performance of external wall made of highly perforated bricks. *Adv. Eng. Softw.* **2017**, *113*, 47–53. [CrossRef]

31. Kruis, J.; Koudelka, T.; Krejčí, T. Efficient computer implementation of coupled hydro-thermo-mechanical analysis. *Math. Comput. Simul.* **2010**, *80*, 1578–1588. [CrossRef]

32. Kočí, V.; Kočí, J.; Maděra, J.; Pavlík, Z.; Gu, X.; Zhang, W.; Černý, R. Thermal and hygric assessment of an inside-insulated brick wall: 2D critical experiment and computational analysis. *J. Build. Phys.* **2018**, *41*, 497–520. [CrossRef]

33. Pavlík, Z.; Fiala, L.; Vejmelková, E.; Černý, R. Application of Effective Media Theory for Determination of Thermal Properties of Hollow Bricks as a Function of Moisture Content. *Int. J. Thermophys.* **2013**, *34*, 894–908. [CrossRef]

34. Čáchová, M.; Koňáková, D.; Vejmelková, E.; Keppert, M.; Polozhiy, K.; Černý, R. Pore Structure and Thermal Characteristics of Clay Bricks. *Adv. Mater. Res.* **2014**, *982*, 104–107. [CrossRef]

35. Vejmelková, E.; Pavlíková, M.; Keppert, M.; Keršner, Z.; Rovnaníková, P.; Ondráček, M.; Sedlmajer, M.; Černý, R. High performance concrete with Czech metakaolin: Experimental analysis of strength, toughness and durability characteristics. *Constr. Build. Mater.* **2010**, *24*, 1404–1411. [CrossRef]

36. Kočí, V.; Maděra, J.; Fořt, J.; Žumár, J.; Pavlíková, M.; Pavlík, Z.; Černý, R. Service Life Assessment of Historical Building Envelopes Constructed Using Different Types of Sandstone: A computational Analysis Based on Experimental Input Data. *Sci. World J.* **2014**, *2014*, 802509. [CrossRef]

37. Jiřičková, M.; Černý, R. Effect of hydrophilic admixtures on moisture and heat transport and storage parameters of mineral wool. *Constr. Build. Mater.* **2006**, *20*, 425–434. [CrossRef]

38. Kočí, V.; Maděra, J.; Jerman, M.; Žumár, J.; Koňáková, D.; Čáchová, M.; Vejmelková, E.; Reiterman, P.; Černý, R. Application of waste ceramic dust as a ready-to-use replacement of cement in lime-cement plasters: An environmental-friendly and energy-efficient solution. *Clean Technol. Environ. Policy* **2016**, *18*, 1725–1733. [CrossRef]

39. Vejmelková, E.; Keppert, M.; Keršner, Z.; Rovnaníková, P.; Černý, R. Mechanical, fracture-mechanical, hydric, thermal, and durability properties of lime-metakaolin plasters for renovation of historical buildings. *Constr. Build. Mater.* **2012**, *31*, 22–28. [CrossRef]

energies

MDPI

Article

Operation Testing of an Advanced Personalized Ventilation System

Imre Csáky, Tünde Kalmár and Ferend Kalmár *

Faculty of Engineering, University of Debrecen, 4032 Debrecen, Hungary; imrecsaky@eng.unideb.hu (I.C.);
kalmar_tk@eng.unideb.hu (T.K.)
* Correspondence: fkalmar@eng.unideb.hu

Received: 5 April 2019; Accepted: 25 April 2019; Published: 26 April 2019

Abstract: Using personalized ventilation systems in office buildings, important energy saving might be obtained, which may improve the indoor air quality and thermal comfort sensation of occupants at the same time. In this paper, the operation testing results of an advanced personalized ventilation system are presented. Eleven different air terminal devices were analyzed. Based on the obtained air velocities and turbulence intensities, one was chosen to perform thermal comfort experiments with subjects. It was shown that, in the case of elevated indoor temperatures, the thermal comfort sensation can be improved considerably. A series of measurements were carried out in order to determine the background noise level and the noise generated by the personalized ventilation system. It was shown that further developments of the air distribution system are needed.

Keywords: advanced personalized ventilation; energy saving; air terminal device; air velocity; turbulence; noise level

1. Introduction

In European countries, buildings account for 40% of the total energy consumption [1]. The sector is expanding and the comfort needs of occupants are increasing. According to Vorsatz et al., the share of heating and cooling in the building energy balance is variable between 18% and 73% [2]. Therefore, energy saving in the building sector is primordial in the European Union. According to 2020 energy goals, the energy efficiency should increase by 20%. The share of renewable energy sources should increase by 20% and the CO2 emissions should be reduced with 20% or even 30%. [3]. Climate change does not help in achieving the goals. It was proven by different scholars that, in European countries, future summers are getting warmer and the number and amplitude of heat waves will increase in the future [4,5]. According to Isaac and van Vuuren, climate change has little net effect on the global energy use because the increase of the outdoor temperature leads to the decrease of heating energy use. However, at the same time, it induces the increase of the cooling energy consumption [6]. However, if the heating and cooling energy are analyzed separately, the impact of climate change is significant. According to their scenario, by 2100, the heating energy needs will decrease by 34%, while the cooling energy needs are going to increase by 72%. Levesque et al. demonstrated that, without new climate policies and drastic changes in the energy use, the global final energy need of buildings could increase from 116 EJ/yr in 2010 to anywhere from 120 to 378 EJ/yr in 2100 [7]. According to their results, buildings' energy demand will be dominated by the energy use of appliances, lighting, and space cooling, while the share of heating and cooking decrease. Santamouris analyzed the energy use of buildings assuming three different scenarios: based on low, average, and high future development [8]. He predicted a range of the expected cooling energy demand in 2050 assuming various boundary conditions. His results show that the average cooling energy demand of the residential and commercial buildings in 2050 will increase up to 750% and 275%, respectively. The decrease of heating share and increase of cooling needs in the building energy balance are influenced by the severe requirements related to thermal properties of the buildings' envelope. New insulating materials are tested in order to meet the requirements related to opaque building elements with thin layers [9,10]. However, in the case of properly insulated buildings, even small heat loads may lead to high indoor air temperatures and the cooling energy need increases while the heating energy demand decreases. To optimize the facade and building shell solutions, such as window areas and physical properties, external wall insulation type and thickness, window-to-wall ratio, and external shading simulations and cost optimal calculations were performed [11]. In the case of free running office or educational buildings with large glazed areas, in spite of the relatively high air change rates provided through natural ventilation, extreme high indoor temperatures may appear [12,13]. Jakubcionis and Carlsson estimated the total potential space cooling demand of the EU to be 292 TW h for the residential sector in an average year [14]. They estimated the additional electrical capacity needed for 79 GW. They stated that, with proper energy system development strategies, the stresses on electricity system from increasing cooling demand can be mitigated.

Air leakage represents an important factor in the energy balance of buildings and influences the indoor air quality as well. Chan et al. analyzed more than 70,000 air leakage measurements in houses across the United States to obtain information about the relation between air leakage area–considered as the effective size of all penetrations of the building shell–and available building characteristics such as building size, year built, geographic region, and various construction characteristics [15]. They presented the regressions of normalized leakage for three house classifications: low-income households, energy program houses, and conventional houses. D'Ambrosio et al. presented the results of a survey on residential buildings located in southern Italy using the fan pressurization method [16]. They found that the average air change rate n50 value is fairly high, particularly for the buildings built before the 1970s. Thus, the air tightness of new buildings is better and the air leakage is reduced considerably. However, Sinnott and Dyer highlighted the importance of workmanship and construction detailing in order to achieve the required air tightness [17]. In the case of new buildings, the improved air tightness of the envelope may lead to the increase of carbon dioxide concentration and humidity content of the indoor air [18]. Low ventilation rates in dwellings increase the risk of allergic symptoms among children [19]. In order to choose the appropriate ventilation strategy in buildings, complex studies have to be performed [20]. Local or personalized ventilation may be the optimal solution providing proper air quality in the occupational zone with minimal energy use. Cao et al. concluded in their study that the combination of different types of ventilation, like displacement ventilation and mixing ventilation, personalized ventilation, and displacement ventilation might have a better performance than using only one method [21]. According to Melikov, the focus must be shifted from total volume air distribution to advanced air distribution based on the following principles [22]:

- involve each occupant in creating his/her own preferred microenvironment;
- remove/reduce the air pollution and generated heat (when not needed) locally;
- make possible active control of the air distribution;
- provide clean air as well as heating and cooling as much as needed in any location and at any point in time.

At the Building Services and Building Engineering Department, University of Debrecen, an advanced personalized ventilation equipment was developed (ALTAIR), which provides the air jet around the head and chest of the occupants alternatively from different directions (left-front-right), [23]. Numerous measurements were carried out and it was proven that ALTAIR improves the thermal comfort of occupants in environments with elevated operative temperatures [24–27]. However, the equipment needs to be improved further.

The air terminal device plays a key role in the operation of personalized ventilation equipment. Conceição et al. drew the attention about the local thermal discomfort conditions, associated to the draught risk index and to the air velocity fluctuation equivalent frequency [28]. Tham and Pantelic concluded in their study that the risk of draft caused by intensive cooling of small areas of the body can be reduced by cooling when it is more evenly distributed across the whole body surface [29]. According to Melikov et al., performance of personalized ventilation systems depends largely on the supply air terminal device (ATD) [30]. They developed, tested, and evaluated five ATD's. They stated that the personal exposure effectiveness increased with the increase of the airflow rate from the ATDs to a constant maximum value, which was not affected by a further increase of the airflow.

Another challenge of the ALTAIR personalized ventilation system is to reduce the noise to an acceptable level. According to Kjellberg, noise is probably the most widespread problem in the physical work environment [31]. He revealed that the noise level is raised in offices because of ventilation systems, computers, printers, and other machines. He drew the attention on noise annoyance and its possible behavioral and physiological consequences. The study of Barclays et al. illustrated the importance of an integrated approach to noise exposure and ventilation performance in urban buildings [32].

The aim of the present paper is to present the results of measurements related to an analysis of different air terminal devices and noise level created by the ALTAIR advanced personalized ventilation equipment.

2. ALTAIR PV Equipment

The ALTAIR advanced personalized ventilation equipment is practically integrated in a desk and provides a possibility to introduce the air around the occupants from three different directions (Figure 1).

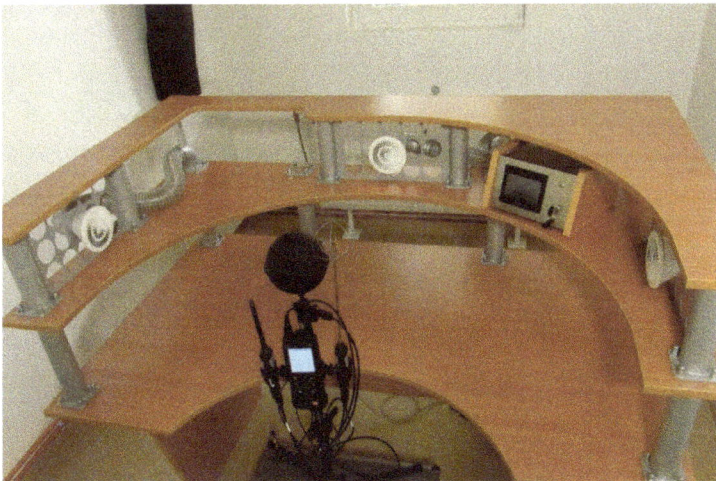

Figure 1. ALTAIR PV equipment.

The ventilated air flow, the air velocity, and the time step of the air jet direction changing is chosen by the occupant. Thus, a comfort bulb is created around the occupant according to her or his needs.

3. Analysis of Air Terminal Devices

In order to improve the air distribution around the occupant and reduce the risk of draught sensation, 11 different air terminal devices were tested. The main analyzed characteristics were the air jet profile, mean air velocity, and turbulence. Taking into account that the fresh air demand depends on the activity level, new testing equipment was built in order to perform measurements on a larger scale of the airflow. The equipment has a Ø160 duct fan with continuous speed control (Figure 2). The air duct diameter was reduced from 160 mm to 100 mm in order to provide the connection possibility for the air terminal devices.

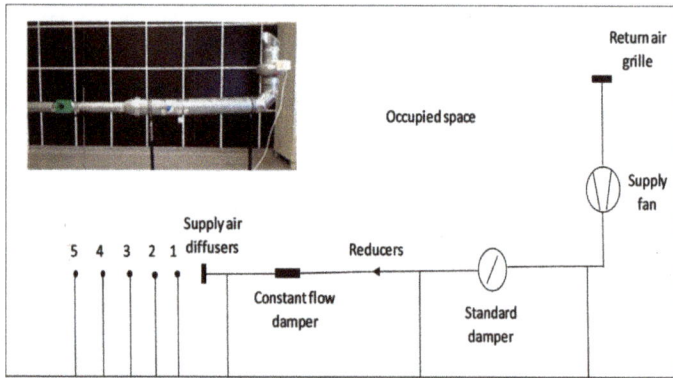

Figure 2. Testing equipment for air terminal devices.

The air velocities and the turbulence were measured with calibrated TESTO 480 instrument at 10, 20, 30, 40, and 50 cm from the ATD's. According to the calibration certificate, the uncertainty of measured velocities is 0.02 m/s (coverage factor k = 2, 95% confidence). The airflow was set to 30, 40, and 50 m³/h. In Figure 3, the air velocities can be seen in the case of the simple air duct with a 100-mm diameter.

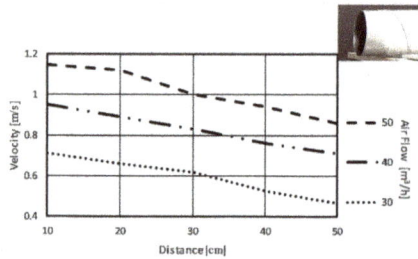

Figure 3. Mean air velocities in the case of D = 100 mm air duct.

Analyzing the results obtained for 11 ATD's, in light of our goals, only five gave acceptable results. The mean air velocities are presented in Figures 4–8.

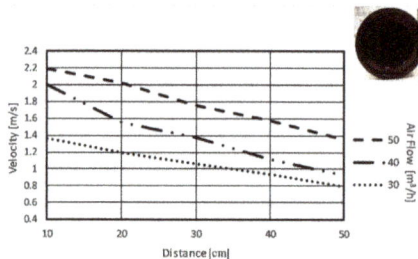

Figure 4. Mean air velocities in the case of the SAR circular perforated ATD.

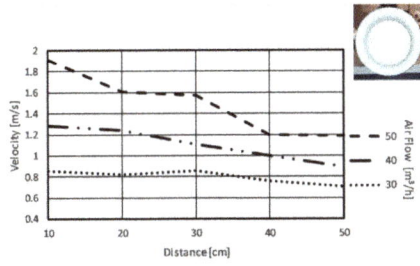

Figure 5. Mean air velocities in the case KV 100 type ATD.

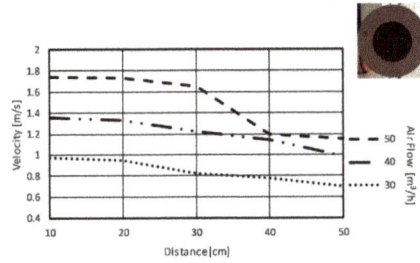

Figure 6. Mean air velocities in the case of E50/100 type ATD.

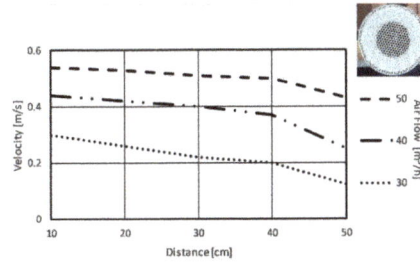

Figure 7. Mean air velocities in the case of TSP 100 type ATD.

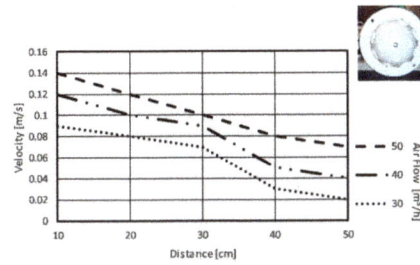

Figure 8. Mean air velocities in the case of closed ZMD type ATD.

The mean turbulence intensities for these ATD types were: 7.52% (50 m³/h), 7.36% (40 m³/h), and 15.56% (30 m³/h). For the other six ATD's, the turbulence was quite high: 68.04% (50 m³/h), 59.52%

(40 m³/h), and 74.72% (30 m³/h). We decided to work with the air terminal devices, which provide low turbulence intensities.

For air flow visualization, smoke tests were carried out (50 m³/h air flow). The results are presented in Figure 9.

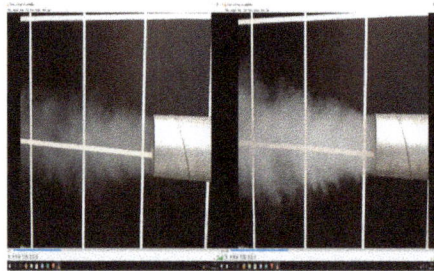

(a)

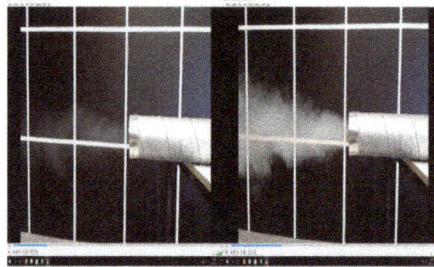

(b)

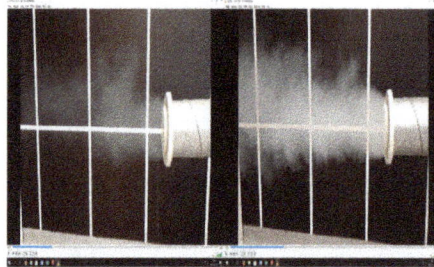

(c)

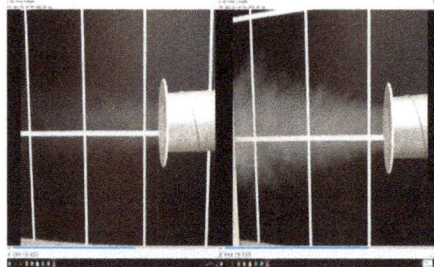

(d)

Figure 9. *Cont.*

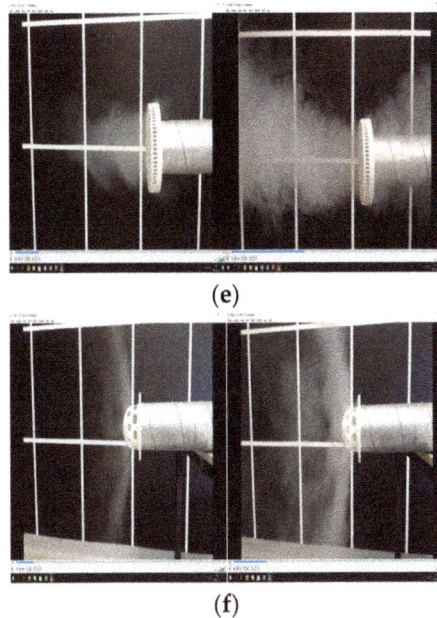

Figure 9. Visualization of air flow with smoke tests: (**a**) D = 100 mm air duct. (**b**) SAR circular perforated ATD, (**c**) KV 100, (**d**) E50/100, (**e**) TSP 100, and (**f**) closed ZMD type ATD.

In the following, the SAR circular perforated ATD was used for experiments.

4. Energy Aspects of ALTAIR PV Equipment

The results of 2.0 h long measurements carried out in a closed space (ALTAIR in operation) with 30 °C air and mean radiant temperatures show that the actual mean vote of male subjects (age: 55.2 ± 3.6 year) was 0.78, while the actual mean vote of female subjects (age: 59.1 ± 3.0 year) was 0.52. The ventilated airflow was 20 m^3/h and the air jet was isothermal. In comparison, providing 50 m^3/h fresh air, with displacement ventilation, assuming activity: 1.2 met, clothing: 0.5 clo, and the PMV measured with TESTO 480 instrument was 1.44. With the ALTAIR PV system in operation, the air velocity was 0.48 m/s (the turbulence was Tu_{10} = 20.6% at 10-s time step of the air jet direction changing, Tu_{20} = 19.1% at 20-s time step and Tu_{30} = 18.8% at 30-s time step). The PMV measured with TESTO 480 instrument was 0.84. The air terminal device was an SAR perforated plastic circular element (D = 75 mm). The distance between ATD's and occupants' head was 0.6 m. The energy consumed by the ALTAIR's fan during a one-hour operation was 12 Wh.

Assuming the occupant in the closed space includes 0.1 m/s air velocity (v_a), 50% relative humidity (RH), and 30 °C mean radiant temperature (MRT), the 0.78 and 0.52 actual mean votes could be obtained providing cool air in the room (Figure 10).

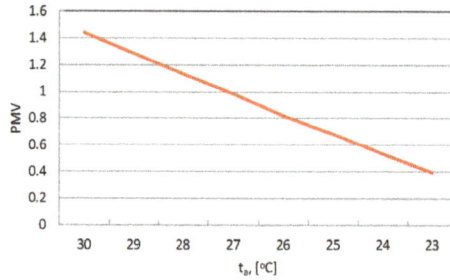

Figure 10. PMV depending on the air temperature(MRT = 30 °C. RH = 50%. v_a = 0.1 m/s).

Assuming cooling with compressors (SEER = 3.6), the energy used in one hour will be 92 Wh and 64 Wh, respectively. It can be observed that important energy savings can be obtained using the ALTAIR PV equipment. However, to reduce as much as possible (or eliminate) the draught sensation and risk and to decrease the noise level, further research was carried out.

5. Noise Measurements

The laboratory building of the faculty of Engineering, University of Debrecen (Figure 11) has a reinforced frame structure and has two levels. The external walls are built from gas silicate blocks (36 cm) and are covered on the external side with 8-cm mineral wool.

Figure 11. Measuring points outside the building (Source: https://www.google.com/maps/place/Debrecen).

The Indoor Environment Quality laboratory is placed at the second level. The height of the floor level is 6.8 m and the height of the second level is 4.5 m. The internal height of the 3.65 m × 2.5 m test room is 2.4 m and is placed in a climate chamber inside the building (Figure 12). Between the test room walls and climate chamber panels, there is a space with the same internal height as the test room has. The width of this space is 1.5 m. The climate chamber is built from 15-cm thick polyurethane panels.

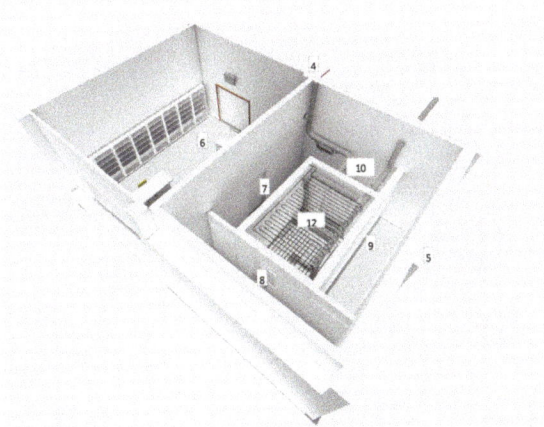

Figure 12. Schema of the Indoor Environment Quality laboratory, University of Debrecen.

The noise level was measured in three points outside the building (Figure 11), in the test room (point 12), and seven points around the test room (Figure 12).

To determine the noise levels, a calibrated TESTO 815 sound level meter was used (Figure 13). The measuring range is +32 to +130 dB and de accuracy is ±1.0 dB. The noise levels measured around the building, around the test room, and in the test room without any ventilation are presented in Figure 14. The duration of one measurement was 1.0 h. The data were gathered every 5.0 min.

Figure 13. Testo 815 sound level meter.

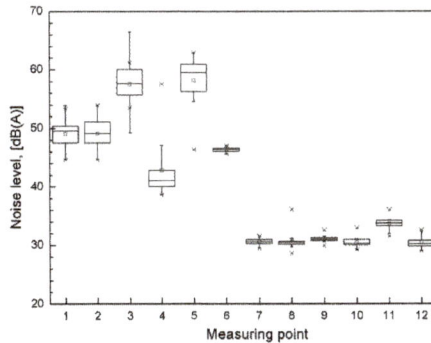

Figure 14. Basic measurements.

It can be observed that, in the climate chamber, the noise level was about 32 to 35 dB(A). In the test room, the ventilation can be realized in mixing mode or displacement mode. The noise level was measured for both types of background ventilation for 50 m³/h ventilated airflow and 500 m³/h ventilated airflow (Figure 15).

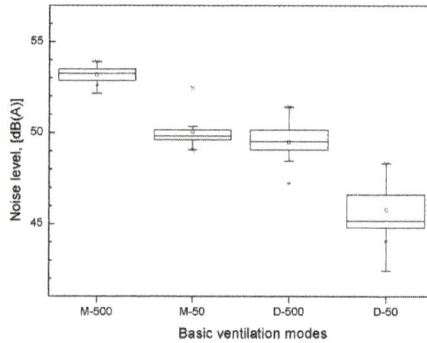

Figure 15. Noise levels in the test room (with background ventilation).

During the measurements with ALTAIR in operation, the 50 m³/h displacement ventilation was chosen as background ventilation (D-50). This ventilation mode leads to 42 to 48 dB(A) noise levels in the test room. The ALTAIR PV system can be connected to the fresh air ducts or can blow on the occupant of the indoor air from the closed space. In the case of our measurements, the air source for ALTAIR PV system was the closed space. Furthermore, the occupant can choose the airflow rate and the time step of the changing air jet direction. It was decided to measure the noise levels at four different airflows: 9.0 m³/h, 18 m³/h, 27 m³/h, and 36 m³/h. The highest airflow is practically 10 l/s (often used in the indoor air quality standards). The time steps of changing the air jet directions were: 10 s, 20 s, and 30 s. The noise levels measured at different time steps for 9.0 m³/h airflow are presented in Figure 16.

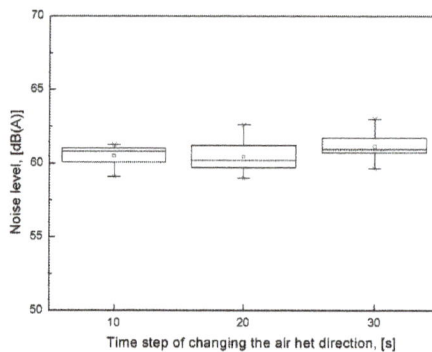

Figure 16. Noise levels in the test room (with background ventilation and ALTAIR in operation—9.0 m³/h).

By measuring different airflows, it was found that changing the time step of the air jet had no effect on the noise level. In the following information, the results for a 10-s time step are shown.

In Figure 17, it is observed that the air flow chosen at the ALTAIR PV system has an important effect on the noise level in the room. The noise level increases from 61 dB(A) to 66 dB(A), if the air flow is risen from 9 m³/h to 36 m³/h.

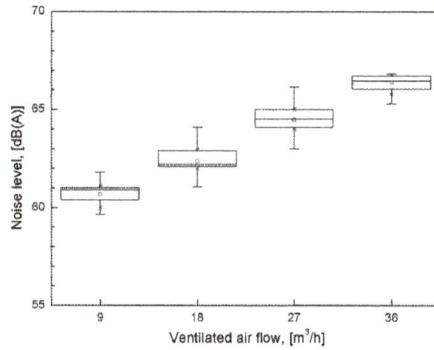

Figure 17. Noise levels in the test room (with background ventilation and ALTAIR in operation—10 s time step).

In order to reduce the noise level in the test room, an Audimin-DW-S sound absorber panel [33] was placed near the fan of the ALTAIR PV system. The measured noise levels are presented in Figure 18.

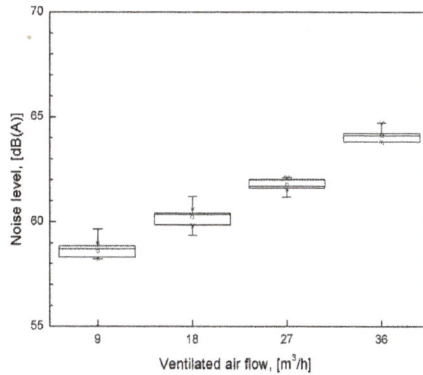

Figure 18. Noise levels in the test room (with background ventilation, ALTAIR with sound absorber—10 s time step).

6. Discussion

The air blown on the chest and head of the occupant besides proper indoor air quality should provide slight cooling sensation without draught perception. After analyzing the air velocities and airflow distributions obtained for different air terminal devices with low turbulence intensities, it was stated that four might be used for the ALTAIR PV system: SAR circular perforated ATD, KV 100, E50/100, and TSP 100. Thermal comfort measurements were performed with SAR circular perforated ATD. In indoor environments with elevated air and mean radiant temperatures (30 °C), the actual mean votes were reduced from 1.44 to anywhere from 0.51 to 0.78. However, 20% to 30% of subjects felt draught (0.48 m/s air velocity and 20% turbulence). Even though the percentage of dissatisfied persons was lower (because at elevated air and mean radiant temperatures draught might improve the thermal comfort), new air terminal devices have to be tested and developed in the future. Practice has shown that the noise level of an air conditioner is about 60 dB(A) [34]. The displacement ventilation in the test room generates a 60 dB(A) – 61 dB(A) noise level. The noise level generated by the ALTAIR PV system with background displacement ventilation (50 m³/h) ranges from 61 dB(A) to 66 dB(A), depending on the air flow chosen by the occupant. Experiments show that these noise levels can be reduced with sound absorbers under 60 dB(A). However, the air distribution system of ALTAIR PV equipment has to be improved in order to further reduce the noise level.

7. Conclusions

The developed ALTAIR PV system improves the thermal comfort sensation of occupants in indoor environments with an elevated mean radiant and air temperatures. At the same time, important cooling energy can be saved in comparison with traditional cooling systems with mechanical compressors. By testing eleven different air terminal devices, it was proven that only five devices provide turbulence intensities below 20%. At elevated air velocities, the low turbulence is indispensable in order to avoid draught sensation. The obtained results show that, in order to obtain the optimal air velocity and flow distribution around the occupants, new air terminal devices have to be developed. Furthermore, special attention has to be paid to the air distribution system of the ALTAIR PV equipment in order to reduce the generated noise level.

8. Limitation

Measurements were performed in the laboratories of the Faculty of Engineering, University of Debrecen. The size of the test room, the building materials, and the surfaces of building elements of the test room as well as the type and location of air terminal devices of mixing and displacement ventilation were considered as given boundary conditions.

Author Contributions: Conceptualization, F.K. Methodology, T.K., I.C., and F.K. Investigation, I.C., T.K., and F.K. Resources, I.C. and F.K. Data curation, I.C. and T.K. Writing—original draft preparation, I.C., T.K., and F.K. Writing—review and editing, F.K. Visualization, I.C. and T.K. Supervision, F.K. Project administration, F.K.

Funding: The Higher Education Institutional Excellence Program of the Ministry of Human Capacities in Hungary, within the framework of the Energetics thematic program of the University of Debrecen, financed the research.

Conflicts of Interest: The authors declare no conflicts of interest.

References

1. Directive 2010/31/EU of the European Parliament and of The Council of 19 May 2010 on the Energy Performance of Buildings. Available online: https://eur-lex.europa.eu/LexUriServ/LexUriServ.do?uri=OJ:L:2010:153:0013:0035:en:PDF (accessed on 18 March 2019).
2. Ürge-Vorsatz, D.; Cabeza, L.F.; Serrano, S.; Barreneche, C.; Petrichenko, K. Heating and cooling energy trends and drivers in buildings. *Renew. Sustain. Energy Review.* **2015**, *41*, 85–98. [CrossRef]
3. RECS International. Available online: http://www.recs.org/glossary/european-20-20-20-targets (accessed on 18 March 2019).
4. Luterbacher, J.; Dietrich, D.; Xoplaki, E.; Grosjean, M.; Wanner, H. European seasonal and annual temperature variability, trends, and extremes since 1500. *Science* **2014**, *303*, 1499–1503. [CrossRef] [PubMed]
5. Schar, C.; Vidale, P.L.; Lüthi, D.; Frei, C.; Haberli, C.; Liniger, M.A.; Appenzeller, C. The role of increasing temperature variability in European summer heatwaves. *Nature* **2004**, *427*, 332–336. [CrossRef] [PubMed]
6. Isaac, M.; van Vuuren, D.P. Modeling global residential sector energy demand for heating and air conditioning in the context of climate change. *Energy Policy* **2009**, *37*, 507–521. [CrossRef]
7. Levesque, A.; Pietzcker, R.C.; Baumstark, L.; De Stercke, S.; Grübler, A.; Luderer, G. How much energy will buildings consume in 2100? A global perspective within a scenario framework. *Energy* **2018**, *148*, 514–527. [CrossRef]
8. Santamouris, M. Cooling the buildings – past, present and future. *Energy Build.* **2016**, *128*, 617–638. [CrossRef]
9. Umberto, B.; Lakatos, Á. Thermal Bridges of Metal Fasteners for Aerogel-enhanced Blankets. *Energy Build.* **2019**, *185*, 307–315.
10. Lakatos, Á. Stability investigations of the thermal insulating performance of aerogel blanket. *Energy Build.* **2019**, *185*, 103–111. [CrossRef]
11. Thalfeldt, M.; Pikas, E.; Kurnitski, J.; Voll, H. Facade design principles for nearly zero energy buildings in a cold climate. *Energy Build.* **2013**, *67*, 309–321. [CrossRef]

12. Kalmár, F. Interrelation between glazing and summer operative temperature in buildings. *Int. Rev. Appl. Sci. Eng.* **2016**, *7*, 53–62. [CrossRef]
13. Kalmár, F. Summer operative temperatures in free running existing buildings with high glazed ratio of the facades. *J. Build. Eng.* **2016**, *6*, 236–242. [CrossRef]
14. Jakubcionis, M.; Carlsson, J. Estimation of European Union residential sector space cooling potential. *Energy Policy* **2017**, *101*, 225–235. [CrossRef]
15. Chan, W.R.; Nazaroff, W.W.; Price, P.N.; Sohn, M.D.; Gadgil, A.J. Analyzing a database of residential air leakage in the United States. *Atmos. Environ.* **2005**, *39*, 3445–3455. [CrossRef]
16. d'Ambrosio Alfano, F.R.; Dell'Isola, M.; Ficco, G.; Tassini, F. Experimental analysis of air tightness in Mediterranean buildings using the fan pressurization method. *Build. Environ.* **2012**, *53*, 16–25. [CrossRef]
17. Sinnott, D.; Dyer, M. Air-tightness field data for dwellings in Ireland. *Build. Environ.* **2012**, *51*, 269–275. [CrossRef]
18. Zemitis, J.; Borodinecs, A.; Frolova, M. Measurements of moisture production caused by various sources. *Energy Build.* **2016**, *127*, 884–891. [CrossRef]
19. Fanger, P.O. What is IAQ? *Indoor Air* **2006**, *16*, 328–334. [CrossRef]
20. Baranova, D.; Sovetnikov, D.; Semashkina, D.; Borodinecs, A. Correlation of energy efficiency and thermal comfort depending on the ventilation strategy. *Procedia Eng.* **2017**, *205*, 503–510. [CrossRef]
21. Cao, G.; Awbi, H.; Yao, R.; Fan, Y.; Sirén, K.; Kosonen, R.; Zhang, J. A review of the performance of different ventilation and airflow distribution systems in buildings. *Build. Environ.* **2014**, *73*, 171–186. [CrossRef]
22. Melikov, A.K. Advanced air distribution: improving health and comfort while reducing energy use. *Indoor Air* **2016**, *26*, 112–124. [CrossRef]
23. Kalmár, F. Innovative method and equipment for personalized ventilation. *Indoor Air* **2015**, *25*, 297–306. [CrossRef] [PubMed]
24. Kalmár, F.; Kalmár, T. Alternative personalized ventilation. *Energy Build.* **2013**, *65*, 37–44. [CrossRef]
25. Kalmár, F. An indoor environment evaluation by gender and age using an advanced personalized ventilation system. *Build. Serv. Eng. Res. Technol.* **2017**, *38*, 505–521. [CrossRef]
26. Kalmár, F. Impact of elevated air velocity on subjective thermal comfort sensation under asymmetric radiation and variable airflow direction. *J. Build. Phys.* **2017**, *42*, 173–193. [CrossRef]
27. Kalmár, F.; Kalmár, T. Study of human response in conditions of surface heating, asymmetric radiation and variable air jet direction. *Energy Build.* **2018**, *179*, 133–143. [CrossRef]
28. Conceiçao, E.Z.E.; Lúcio, M.M.J.R.; Rosa, S.P.; Custódio, A.L.V.; Andrade, R.L.; Meira, M.J.P.A. Evaluation of comfort level in desks equipped with two personalized ventilation systems in slightly warm environments. *Build. Environ.* **2010**, *45*, 601–609. [CrossRef]
29. Tham, K.W.; Pantelic, J. Performance evaluation of the coupling of a desktop personalized ventilation air terminal device and desk mounted fans. *Build. Environ.* **2010**, *45*, 1941–1950. [CrossRef]
30. Melikov, A.K.; Cermak, R.; Majer, M. Personalized ventilation: evaluation of different air terminal devices. *Energy Build.* **2002**, *34*, 829–836. [CrossRef]
31. Kjellberg, A. Subjective, behavioral and psychophysiological effects of noise. *Scand. J. Work Environ. Health* **1990**, *16*, 29–38. [CrossRef]
32. Barclay, M.; Kang, J.; Sharples, S. Combining noise mapping and ventilation performance for non-domestic buildings in an urban area. *Build. Environ.* **2012**, *52*, 68–76. [CrossRef]
33. Schako. Available online: http://audimin.schako.cz/Audimin.pdf (accessed on 18 March 2019).
34. Noise Help. Available online: https://www.noisehelp.com/noise-level-chart.html (accessed on 18 March 2019).

energies

MDPI

Article

Determination of a Methodology to Derive Correlations Between Window Opening Mass Flow Rate and Wind Conditions Based on CFD Results

Panagiotis Stamatopoulos, Panagiotis Drosatos, Nikos Nikolopoulos and Dimitrios Rakopoulos *

Chemical Process and Energy Resources Institute, Centre for Research and Technology Hellas, Thermi, 57001 Thessaloniki, Greece; stamatopoulos@certh.gr (P.S.); drosatos@lignite.gr (P.D.); n.nikolopoulos@certh.gr (N.N.)
* Correspondence: rakopoulos@certh.gr or dimracop@central.ntua.gr; Tel.: +30-211-106-9509

Received: 28 March 2019; Accepted: 24 April 2019; Published: 26 April 2019

Abstract: This paper presents a methodology for the development of an empirical equation which can provide the air mass flow rate imposed by single-sided wind-driven ventilation of a room, as a function of external wind speed and direction, using the results from Computational Fluid Dynamics (CFD) simulations. The proposed methodology is useful for a wide spectrum of applications, in which no access to experimental data or conduction of several CFD runs is possible, deriving a simple expression of natural ventilation rate, which can be further used for energy analysis of complicated building geometries in 0-D models or in object-oriented software codes. The developed computational model simulates a building, which belongs to Rheinisch-Westfälische Technische Hochschule (RWTH, Aachen University, Aachen, Germany) and its surrounding environment. A tilted window represents the opening that allows the ventilation of the adjacent room with fresh air. The derived data from the CFD simulations for the air mass flow were fitted with a Gaussian function in order to achieve the development of an empirical equation. The numerical simulations have been conducted using the Ansys Fluent v15.0® software package. In this work, the k-w Shear Stress Transport (SST) model was implemented for the simulation of turbulence, while the Boussinesq approximation was used for the simulation of the buoyancy forces. The coefficient of determination R^2 of the curve is in the range of 0.84–0.95, depending on the wind speed. This function can provide the mass flow rate through the open window of the investigated building and subsequently the ventilation rate of the adjacent room in air speed range from 2.5 m/s to 16 m/s without the necessity of further numerical simulations.

Keywords: natural ventilation; single-sided; CFD; mass flow rate prediction; correlation function

---

1. Introduction

Nowadays, natural ventilation of a building is critical in order to reduce energy consumption for space conditioning (cooling and ventilation). Natural ventilation is the process by which clean air, normally outdoor air, is intentionally provided to a space and stale air is removed, without using mechanical systems [1]. The purpose of natural ventilation is to achieve maximum human comfort in indoor spaces by ensuring maximum use of natural resources [2].

There are two types of natural ventilation in buildings: wind-driven and buoyancy- driven ventilation [3]. Wind-driven ventilation arises from the different pressure created by external wind conditions, while openings in the perimeter of the building permit the flow infiltration [4]. Buoyancy-driven ventilation occurs because of temperature difference between the interior and exterior air. Normally, there is a combination of these two phenomena of ventilation. In order for the designers to improve the energy efficiency of the buildings, they use two mechanisms for improving the natural ventilation: (a) the single-sided ventilation (SS) and (b) the cross-flow ventilation (CR) [5,6]. In both cases, ventilation is either wind or buoyancy-driven or both and occurs through the openings of the building. Cross-flow ventilation (CR) is achieved using windows on both sides of a building. SS ventilation describes a situation in which wind enters and leaves the building through one or two openings located on the same side of the building or the room. Ventilation within the building is mainly affected by the geophysical morphology and the surrounding buildings [7], the used building side for the ventilation mechanism, the type of the openings and the external wind conditions [8]. Therefore, the numerical study of the natural ventilation process is complicated due to the fact that the airflow is affected from multiple factors simultaneously.

As understood, SS ventilation uses only one side of the building, so it is less efficient compared to CR ventilation. Although single-sided ventilation is less efficient, it is more suitable for cellular room environments, such as offices, because they have not openings on both wall sides and cannot implement the cross-flow ventilation mechanism [9].

In general, there are two main methods to define the mass flow rate and the flow parameters through a building's opening; the first one is the calculation by experiments either in full-scale buildings or in wind tunnel [10] and the second one is the numerical prediction [11]. The numerical approach gives the flexibility to study several cases of building structures and environmental conditions. In this paper, our efforts are focused on the development of a function that can predict the ventilation air mass flow rate through a building's opening as a function of external wind direction and speed.

Several investigations have been carried out in order to develop empirical equations that are capable of predicting the ventilation rate of buildings. In 1980, Phaff and De Gids et al. [12] proposed an empirical expression to calculate the airflow rate in a single-sided ventilation zone, based on opening area, wind speed and air temperature. The empirical expression is based on measurements that have been performed on the first floor of buildings, which are surrounded by other buildings up to four floors high.

Warren and Parkins [13,14] also proposed an empirical expression for buoyancy-driven and one for wind-driven single-sided natural ventilation. These expressions are function of the opening dimensions, such as window's area and height, the gravitational acceleration, the average temperature difference between indoor spaces and outdoor environment and wind speed. A combination of these two expressions by quadrature function yields the final equation for the calculation of ventilation flow rate. In 2005, the American Society of Heat, Refrigerating and Air-Conditioning Engineers [15] proposed the first mathematical expression to calculate airflow in single-sided ventilation that takes into account the wind angle of incidence.

Many investigations used the empirical expressions that have already been mentioned for validation purposes. In this framework, Alloca et al. [16] investigated the single-sided natural ventilation of a building by using a CFD model together with an analytical/empirical model. The analytical method involves the equations from Bernoulli theory for the buoyancy-driven flow with an empirical discharge coefficient using the empirical model of Phaff and de Grids [12], for single-sided ventilation. This computational study investigates two cases. The first case is a typical student dormitory with two openings; an upper and a lower window on the same wall side. The second case is a three-story apartment composed of three identical dormitory rooms stacked vertically above one another. The CFD model follows the same trends as the empirical model, but underestimates the ventilation rates by 35%.

Asfour et al. [17] presented a comparison between the airflow rate calculated with a network mathematical model and the airflow rate calculated with CFD simulations. In this work, three different rooms with nearly the same volume, but different aspect ratios, were studied. Each case examines two wind speeds, 1 and 5 m/s, and two wind angles, namely 0° and 45°. The discrepancy percentage between estimated and calculated airflow rate is in the range of −11.5% and 5.3%. Due to the good agreement between the results of the two models, the network mathematical model can be used as a validation tool of CFD studies that have no access to experimental data.

Caciolo et al. [18] examined the accuracy of the air change rate predictions by the already existing empirical correlations of Warren, Phaff and De Gibs [12], Larsen and Heiselberg [19] and

Dascalakis [20]. They conclude that in case of leeward opening, all correlations overestimate the airflow rate. This is attributed to the fact that these correlations do not consider the reduction of the stack effect due to the existence of turbulent diffusion at the opening. On the contrary, in the case of windward opening, the Warren's correlation shows the best agreement with experiments. In a later investigation, Caciolo et al. [21] presented a new correlation for the case of the leeward side by using CFD simulation results. The new correlation shows a better agreement with experiment results, compared to existing correlations.

Tang et al. [22] proposed a new hypothetical correlation based on an experimental study for prediction of airflow rates in a primary school in Beijing in the case of low air speed values and insignificant buoyancy effects, which implement the development of unorganized airflow. Conducting 168 h of experiments, they compare the measured airflow rates against the values derived from the already existing correlations [12,14,19,20]. In a second step, they propose a new hypothetical correlation capable of predicting more accurately the airflow rates in the case of unorganized airflow. The new correlation shows a good prediction of airflow rate. The average deviation is reduced to 17.37%, which is 7% less than the lowest deviation attained from existing correlations.

Wang et al. [23] presented an empirical model capable of predicting the mass flow rate induced by single-sided wind-driven ventilation due to the pressure difference along an opening height. For validation purposes, CFD simulations are performed. The maximum difference between the empirical model prediction and the CFD results is less than 25%. The largest difference is found in the case of leeward side, in which the flow field near the opening is much more complicated. In a later investigation, Wang [24] studied the impact of three types of window, i.e., hopper, awning and casement, in the case of SS natural ventilation. The expression of airflow rate is a function of window and building geometry, opening angle, wind incident angle and speed. These semi-empirical models are based on the previously analytical model and on pressure coefficients. The validation of these expressions has been achieved by using experimental measurements with the tracer-gas method combined with CFD numerical simulations. The new semi-empirical model for predicting the aeration rate for the three types of window presents a good agreement between the measured data and the CFD results.

Pan et al. [25] presented a model for calculating the ventilation rate in SS natural ventilation of an apartment due to wind- and buoyancy-driven effects, based on their previous empirical model [23]. The model is validated by using measured data and is able to predict natural ventilation rate with an average error of 12.7%. The air temperature difference between the indoor and outdoor space is ranged from −2.3 K to 13.2 K. Compared to other six empirical correlations [12,13,18–20,22], this model performs well due to the fact that the other models calculate the ventilation rates with average errors ranging from 12.9% to 46.1%. Moreover, this model takes into account the impact of both positive and negative buoyancy forces along with outside air pressure on natural ventilation through a single opening in contrast with the other models available.

As it was expected, the preceding literature survey shows that none of the models available in the literature is ideal. Although the existing models are sophisticated and have functional forms, the majority of them requires the knowledge of pressure coefficients, discharge coefficients or correction factors for each type of opening. These coefficients are usually obtained experimentally or from standard pressure coefficients. Therefore, it seems that there is an obvious gap in the literature regarding the existence of a simple, but versatile and credible, methodology for the derivation of an expression, which can provide the aeration rate through a building's opening with no dependence on sophisticated experimental or numerical data. Furthermore, this work has examined a wider range of wind speeds compared to existing studies in order to derive the mathematical function of the airflow rate. The selected wind speeds are equal to 6, 10 and 15 m/s, which correspond to 2, 5 and 7 bft wind speeds on the Beaufort scale, respectively. Furthermore, there is no symmetry in the model, in contrast to other studies, since this is a real building and not a theoretical one. Finally, for each case of wind speed, five different wind directions have been applied.

The implemented model has already been developed and validated against experimental data in a well-controlled environment inside a wind tunnel by some members of the present research group in past studies [26–28]. However, the direct comparison of the CFD results against the experimental data regarding the aeration rate of the examined room is not possible, since the wind speed and direction that are used as time-averaged input values in the CFD model are measured by a weather station that is located on the top of the building inside the developed boundary layer. Therefore, due to recirculations inside the boundary layer, these values cannot be considered as representative to time-averaged values of the free airflow conditions and the exerted aeration rates of the investigated room cannot be supposed that have been resulted by these wind conditions. More information regarding the problem with the location of the weather station and the experimental data will be further provided in the Results section. Moreover, an extra effort of validity has been made using previous empirical correlations. However, this effort was not successful, because the existing empirical correlations include pressure, discharge and correction coefficients. These parameters are case-dependent, significantly affected by the opening type and usually defined by experiments. In this specific work, there is no available experimental data for these specific coefficients, so the implementation of the respective equations is not possible either. Additionally, the semi-empirical model of Wang includes the term of neutral plane (plane with zero air velocity), which cannot be defined in this work, due to the type and the opening of the window. Therefore, it was also impossible to make a direct comparison of the results provided by this work with this specific model.

In order to achieve accurate simulation of air-flow, a three-dimensional model has been chosen, based on a building that is located at RWTH Aachen University. An appropriate expression for aeration rate as a function of wind speed and angle of incidence is fitted to the normalized data. In this study, the best fit was achieved using a type of Gaussian function. In order to ensure the verification of the derived equation, three additional wind speeds have been selected; one inside the range of 6 to 15 m/s and two outside the limits of this range, to compare the calculated airflow rates against the estimated ones by the mathematical function. The agreement is good since the maximum relative difference is below 10%, except the cases with wind flow parallel to the building, where the maximum relative difference can be as high as 38%. This high relative difference is attributed to recirculations that are developed in front of the window opening and the empirical correlation cannot take into account. More information regarding this issue can be found in the Results section. To sum up, this methodology is useful for a wide spectrum of applications, in which no access to experimental data or conduction of several CFD runs is possible. Moreover, with this methodology a simple expression of natural ventilation rate can be exported. This expression can be used for further energy analysis of complicated building geometries in 0-D models or in object-oriented software codes. Finally, even if the obtained correlation is not general and can only be used for this specific window type and this specific building envelope, the methodology is generic and can be followed in all cases.

2. Mathematical Model

This work simulates: (a) the developed flow field affected by the wind conditions and the natural convection, (b) turbulence and (c) the energy transfer due to convection and diffusion. The natural convection mechanism is attributed to the implemented temperature difference between ambient air and the room wall temperature. In this work, the applied buoyancy forces are calculated by using the Boussinesq approximation. In combined radiation and convection heat transfer problems, the Boltzmann number represents the ratio of convection to radiation heat transfer [29]. In this work, this specific ratio is very high, i.e., equal to Bo = 334, and thus radiation effects can be neglected. All numerical simulations are solved in steady-state conditions, assuming that the implemented free airflow conditions represent time-averaged values. Since the free airflow conditions (i.e., wind speed and wind direction) are considered to be steady and representative to time-averaged values, it is expected that the transient analysis would eventually have provided the same results with the steady-state analysis regarding the mass flow rate through the window opening, when the calculation time has sufficiently flowed. Therefore, in order to save significant calculation time and avoid convergence issues that might arise in the transient calculation, the steady-state approximation has been followed. Furthermore, since all CFD simulations were steady-state cases and the mesh in tilted window area has high skewness value, the pressure-velocity coupling is achieved by using the Semi-Implicit Method (SIMPLE) algorithm [30,31]. The momentum, energy, turbulent kinetic energy and specific dissipation rate for the first 800 iterations are spatially discretized by using the first-order upwind scheme. After 800 iterations, second-order accuracy schemes are used for momentum and energy equations. The convergence of the steady-state simulation is controlled by monitoring the mass flow rate through the opening using a User-Defined Function (UDF). Figure 1a presents the convergence of the calculated air mass flow rate that gets into the room during the simulation process, assuming 6m/s wind speed and all five cases for the angle of incidence. The simulation of each case is considered converged when the mass flow rate through the opening tend to oscillate around a constant value, see Figure 1b. The residuals of continuity, velocity and energy are below 10^{-5}, 10^{-5} and 10^{-8}, respectively. The mass flow rate for each case is defined by its mean value during the last 500 iterations.

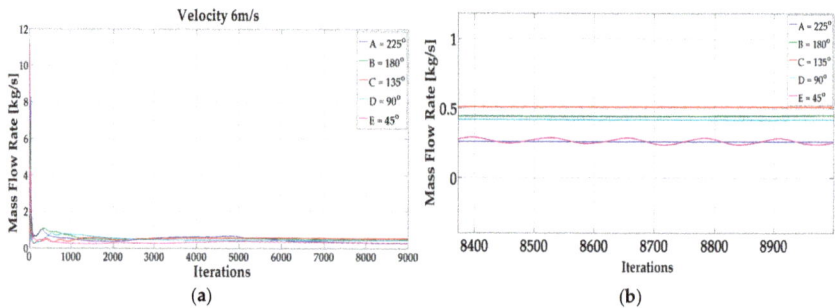

Figure 1. Convergence of air mass flow rate: (**a**) Monitoring of mass flow rate of air that gets into the room, (**b**) oscillations of the mass flow rate.

The general form of continuity equation and the conservation of momentum and energy are given by the Equations (1)–(3), respectively. The transient terms are taken out of the equations, since the numerical simulations are solved in steady-state conditions:

$$\rho\left(\nabla\cdot\vec{v}\right) = 0 \tag{1}$$

$$\rho\nabla\cdot\left(\vec{v}\vec{v}\right) = -\nabla p + \nabla\cdot\bar{\bar{\tau}} + \rho\vec{g} \tag{2}$$

$$\nabla\cdot\left[\vec{v}\left(\rho E + p\right)\right] = \nabla\cdot\left(k_{\text{reff}}\nabla T + \bar{\bar{\tau}}_{\text{reff}}\cdot\vec{v}\right) \tag{3}$$

where p is static pressure, $\bar{\bar{\tau}}$ is the stress tensor and $\rho\vec{g}$ is the gravitational body force. In this study, the term of gravitational body force, $\rho\vec{g}$, contains the Boussinesq approximation (Equation (6)). In Equation (3), T denotes the temperature and k_{eff} is the effective conductivity.

2.1. Turbulence Model

The advantage of the k-w model over the k-epsilon model [32] is the improved performance for the approximation of the boundary layers under adverse pressure gradients and the more accurate predictions regarding: (a) internal flows, (b) flows that exhibit strong curvature, (c) separated flows, and d) jets. However, k-w model also presents a major disadvantage. More specifically, the boundary layer computations are very sensitive to the values of ω in the free stream. In order to overcome this restriction, it is necessary to use the Shear-Stress-Transport (SST) model. Furthermore, Hooff et al. [33] and Ramponi and Blocken reported in [34] that Shear-Stress-Transport (SST) k-ω model provides high accuracy in predicting wind profiles. Thus, the turbulence of the flow was modelled using in all cases the two-equation Shear-Stress-Transport (SST) k-ω based model, developed by Menter [35]. The turbulence kinetic energy, k, and the specific dissipation rate, ω, are obtained from Equations (4) and (5), respectively:

$$\frac{\partial}{\partial t}(\rho k) + \frac{\partial}{\partial x_i}(\rho k v_i) = \frac{\partial}{\partial x_j}\left(\Gamma_k \frac{\partial k}{\partial x_j}\right) + \tilde{G}_k - Y_k + S_k \tag{4}$$

$$\frac{\partial}{\partial t}(\rho \omega) + \frac{\partial}{\partial x_i}(\rho \omega v_i) = \frac{\partial}{\partial x_j}\left(\Gamma_\omega \frac{\partial \omega}{\partial x_j}\right) + G_\omega - Y_\omega + S_\omega + D_\omega \tag{5}$$

2.2. Boussinesq Model

In natural-convection flow problems, Boussinesq approximation provides a faster convergence of the solution procedure compared to the default case where fluid density is considered as a function of temperature. Therefore, this model has also been implemented in this specific case. This model treats density as a constant value in all solved equations, except for the buoyancy term in the momentum equation, Equation (6), which provides the volume forces due to buoyance:

$$\rho - \rho_{ref} = -\rho_{ref}\beta(T - T_{ref}) \tag{6}$$

where ρ_{ref} is the constant density of the flow equal to 1.16 kg/m^3, T_{ref} represents the buoyancy reference temperature, i.e., 286.88 K, and $\beta = 1/T_{ref}$ is the thermal expansion coefficient equal to 0.00343 K^{-1}. The Boussinesq approximation is valid when the product $\beta(T - T_{ref})$ is lower than unity [36]. Thus, as the room temperature T is equal to 292.38 K, the condition is fulfilled and the Boussinesq approximation can be used.

2.3. Normalization

In order to derive the necessary correlations, it is also necessary to define the appropriate dimensionless parameters. The first parameter refers to the normalized mass flow rate and the second one to the dimensionless direction (angle relative to North-to-South direction). The mathematical formulas for the definition of these dimensionless quantities are given by Equations (7) and (8), respectively:

$$\hat{m} = \frac{\dot{m}}{\dot{m}_{max}} \tag{7}$$

$$\hat{\theta} = \frac{\theta - 135^\circ}{180^\circ} \tag{8}$$

where $\dot{m}_{max}$ is the maximum mass flow rate (kg/s) among the simulations belonging to the group of the same wind speed and different wind directions (as already explained), $\dot{m}$ is the actual mass flow rate (kg/s) numerically calculated at the window opening for each specific case and θ is the angle (o) defining the free airflow direction relative to North-to-South one, resulting in the normalized angles of −0.5, −0.25, 0, 0.25 and 0.45 for 45°, 90°, 135°, 180° and 215° wind direction, respectively.

2.4. Geometric Model

The geometric model for the computational simulations has been developed in ANSYS DesignModeler®. The dimensions of the building are 17.98 m × 70.73 m × 11.64 m (L × W × H). According to the existing best practice guidelines of Franke et al. [37], the domain in the flow direction must be extruded by at least eight times the height of the building, when the flow profiles are not available and the flow is blocked to a large extent (e.g., 10%). In this work, the domain has been extruded by approximately 15.5 times the height of the building (H) along z-axis, since the building blockage is quite high (16%). Furthermore, based on the best practices proposed for the case of a single building, the distance between the top of the computational domain and the roof of the building must be 4–10 times the height of the building. In this work, the extrusion of the domain was 2.72 times the height of the building. Even if the distance is lower than the proposed range, it has been ensured, based on the results, that this distance does not insert an artificial acceleration of the flow over the building. Finally, the distance between the building's sidewalls and the lateral boundaries of the computational domain has been selected to be equal to approximately three times the height of the building envelope, since the blockage is quite high (16%). Therefore, the total length of the domain is 197.87 m (z-axis), the total width 142.69 m (x-axis) and the total height 31.64 m (y-axis). The extrusion of the domain far from the building envelop is necessary to simulate the fully developed flow field and to ensure that this is only dependent on the imposed boundary conditions (BCs) and not affected by the building. The extrusion of the domain far from the building envelop is necessary to simulate the fully developed flow field and to ensure that this is only dependent on the imposed boundary conditions and not affected by the building. Figures 2 and 3 present the developed geometric model for the conduction of the numerical simulations. The circle and bold line are used to define the north direction.

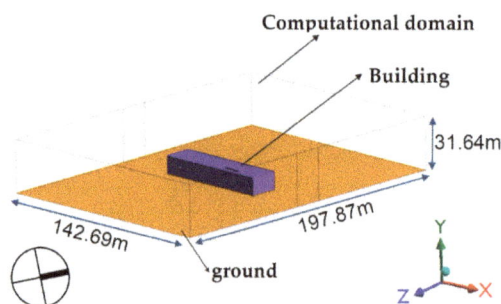

Figure 2. Computational domain.

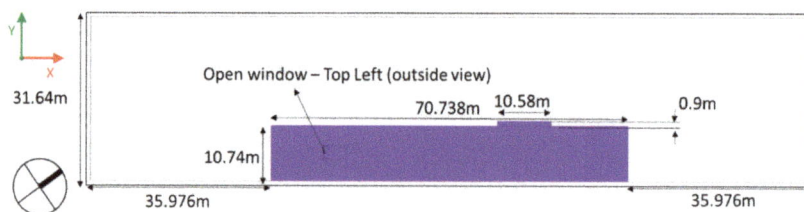

Figure 3. Computational domain (side view).

In addition, the numerical domain includes the investigated room with the specific window opening. The dimensions of the room and the location of the window are shown in Figure 4a. The window itself includes a solid boundary that represents the glass, while its frame is included in the wall boundaries of the building envelop. Furthermore, the opening angle is equal to 5.8° and the effective flow area is equal to 0.485 m^2 (red region in Figure 4a,b). Figure 4b presents the replication of the window opening. The black color represents the window glass, the grey the building walls and the red the window opening. Figure 4c presents the dimensions of the window opening. This type of window was selected as the actual type of the window opening under investigation. In addition, sloping windows are usually found in European homes allowing the efficient ventilation of the building regardless of the weather conditions. The methodology presented in this paper is replicable though, since it can be followed also for other types of windows: awning windows, horizontal and vertical pivot windows or turn windows.

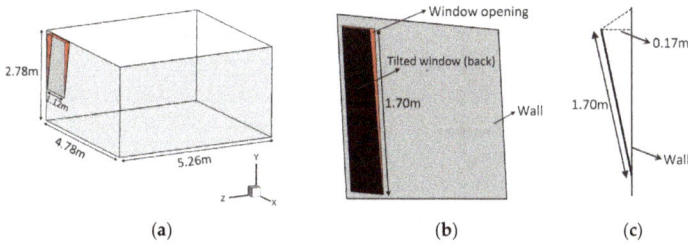

Figure 4. Replication and dimensions of: (**a**) the room, (**b**) window and (**c**) window's opening.

2.5. Boundary Conditions

The bottom surface is defined as a stationary wall with no-slip condition. On the upper surface a symmetry boundary condition is applied. Moreover, a symmetry boundary assumes that the normal velocity and the normal gradients of all variables are equal to zero. The bottom surface (ground) and the interior and exterior walls of the building are modeled as adiabatic walls with no-slip condition. Ambient temperature is defined as equal to 286.88 K and the room walls have a constant temperature equal to 292.38 K. The number and the location of the inlet velocity surfaces of the computational domain are dependent on the wind direction. In general, for every wind speed studied, five different incidence angles are considered, i.e., 45 °, 90 °, 135 °, 180 ° and 215° (Figure 5). In this study, the northwestern wind directions are not examined since transient phenomena of recirculations appear in front of the window area. The angle of incidence of the air flow is relative to the angle defined by North-to-South direction. Therefore, the north wind presents 0° angle of incidence, while the south wind 180°. The outlet surfaces of the computational domain are considered as pressure outlet. The rest surfaces in each case have symmetry boundary conditions. The implemented BCs regarding these crucial operating parameters are given in detail in Table 1.

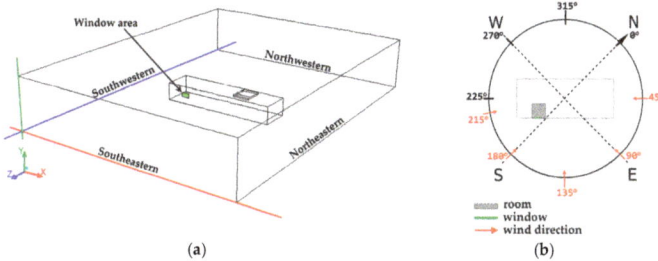

Figure 5. Illustrations of: (**a**) orientation of computational domain, (**b**) top view of building and angle of wind direction.

Table 1. Boundary conditions dependent on the incident angle.

Case	Angle (°)	Velocity Inlet	Pressure Outlet	Symmetry
A	45°	Northeastern	Southwestern	Northwestern, Southeastern
B	90°	Northeastern, Southeastern	Northwestern, Southwestern	-
C	135°	Southeastern	Northwestern	Northeastern, Southwestern
D	180°	Southeastern, Southwestern	Northeastern, Northwestern	-
E	215°	Southwestern	Northeastern	Northwestern, Southeastern

According to Aachen meteorological data, the average wind speed over the course of the year varies in a range of 2 to 16 m/s [38,39]. The mathematical correlation has been derived by using the CFD results in the cases of three different wind speeds, i.e., 6, 10 and 15 m/s. These three values have been commonly agreed with RWTH Aachen University, since for these specific cases the experimental values of aeration rates were available. Nevertheless, as explained in the results section below, the experimental data cannot be used for validation of the CFD model. As a further step, it was necessary to test how accurately the mathematical correlation can predict the aeration rate for any additional wind speed. Therefore, three more wind speeds were used to check the agreement of the CFD results with the results of the empirical expression as regards the mass flow rate through the window opening. For this purpose, three additional wind speeds have been selected; the first is close to the lower bound of the wind range that is typical for Aachen Region, the second one is an intermediate value and the third one is the upper bound of the provided wind range. The selection of the wind directions is arbitrary, except 215°, which is selected due to the fact that for this wind direction the RWTH Aachen University has some experimental values of aeration rates. Nevertheless, as already mentioned, the experimental data cannot be used for validation of the CFD model.

2.6. Numerical Grid

The numerical grid is developed in ANSYS Meshing®. The mesh consists of approximately 8.7 million cells, all of which are hexahedrons. This type of mesh elements can provide smooth solution convergence and validity of the derived results, as compared to the experimental values or the real operating conditions. Mesh shows high quality, since the skewness factor does not exceed in any case the upper acceptable limit of 0.94. Figure 6 presents the developed numerical grid. More specifically, Figure 6a is a general view of the whole domain, while Figure 6b is an enlarged view of the room, showing the numerical grid among the external environment, the room and the tilted window.

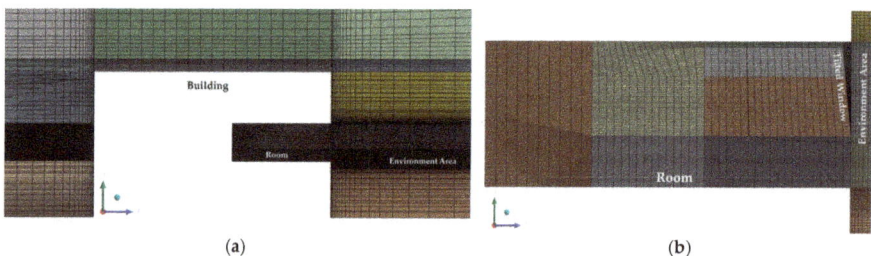

(a)   (b)

Figure 6. Computational mesh of: (**a**) entire domain, (**b**) area between external environment, room and tilted window.

3. Results

3.1. Correlation

In the current study is presented the optimal number of simulations in order to develop the mass flow rate function, since after an assiduous investigation there is no significant effect on the type or coefficients of the fitted functions. Figure 7a presents the CFD results' absolute values of real mass flow rate and wind direction for each case, while Figure 7b presents the respective normalized values. Due to the above explained, mass flow rate normalization formula used for Figure 7b, the mass flow rate normalized values cannot be used to compare the flow in a specific angle among the various wind speeds. For instance, at the angle of 45° (−0.5 of normalized angle), the mass flow rate corresponding to the velocity of 6 m/s is greater than the other two cases, while in the CFD results of Figure 7a, it has the lowest value of the three. In a general view, it can be seen that the mass flow rate follows the same trend for the range between 90° and 180° wind direction. More specifically, the maximum mass flow rate in each case of wind direction is observed in the case with the maximum wind speed, while the minimum one in the case with the minimum wind speed. This is normal, since the mass flow rate is proportional to velocity. However, the case that correspondingly represents the lower bound of the wind direction's range does not follow the same trend. More specifically, in the case with the wind direction of 45°, it can be observed that the mass flow rate in the case with 15 m/s wind speed is located between the respective mass flow values in the cases with 10 m/s and 6 m/s wind speed. In addition, in wind direction of 215°, the cases with 10 m/s and 15 m/s wind speed have almost the same mass flow rate, in contrast to the differences that are detected between these two cases in other wind directions. Finally, the maximum mass flow rate is presented in the case with wind direction of 135°, since in this case the air/mass flow is perpendicular to the window opening, so the normal component of the velocity vector takes the maximum value.

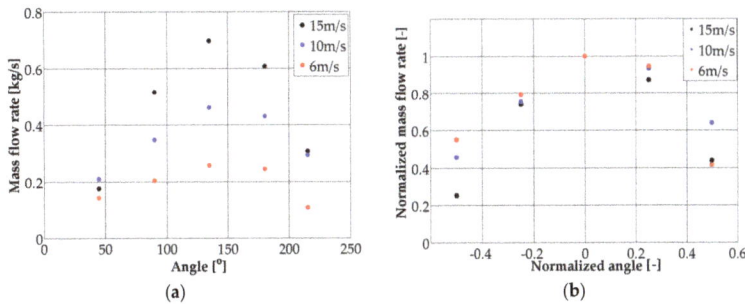

Figure 7. Computational Fluid Dynamics (CFD) results for mass flow rate versus wind direction: (**a**) absolute values, (**b**) normalized values.

The wind conditions were measured by a weather station that is located on the building's roof. However, in Figure 8, case of 135° with 10 m/s wind speed, it can be observed (view of the building from southwestern) that the location is inside the developed boundary layer, so the measurements cannot be considered as representative for time-averaged wind conditions and the measurements of aeration rate as accurate enough for validation purposes.

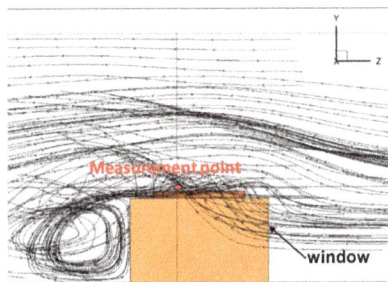

Figure 8. Measurement point and recirculations.

Figures 9–11 show the flow patterns and the horizontal profile of the z-component of velocity in the case of 45° wind direction at a height of 6 m along the y-axis for 6, 10 and 15 m/s wind speed, respectively. These figures show both a general view of the building (a) and a magnified view of the area of interest in front of the window (b). The negative values of velocity z-component represent the direction of the flow towards the window.

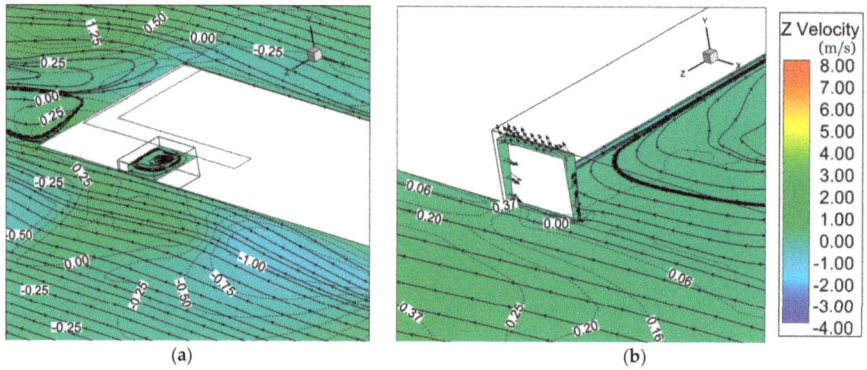

Figure 9. Flow patterns and contour of velocity (z-component), for 6 m/s wind speed and wind direction of 45° (**a**) around the building and (**b**) in front of the tilted window.

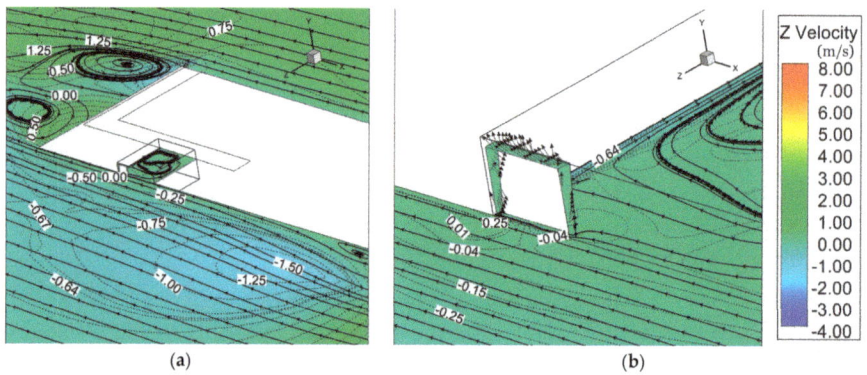

Figure 10. Flow patterns and contour of velocity (z-component), for 10 m/s wind speed and wind direction of 45° (**a**) around the building and (**b**) in front of the tilted window.

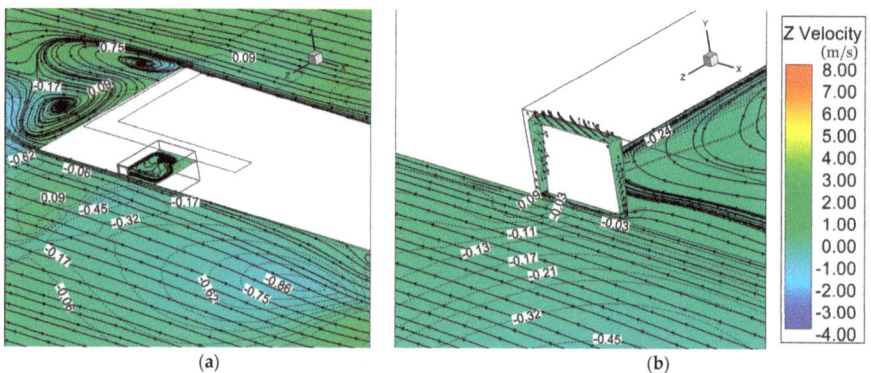

Figure 11. Flow patterns and contour of velocity (z-component), for 15 m/s wind speed and wind direction of 45° (**a**) around the building and (**b**) in front of the tilted window.

Based on the general view of flow patterns around the building, a large recirculation area (kidney vortex) at the leeward side of the building can be observed in all cases examined. As the velocity is increased, the vortices become more intense. Inside the room a clockwise vortex is developed as a result of the closed door. More specifically, due to the closed door, mass flow rates through the window opening are equal. This mutual and simultaneous mass exchange between the external and internal environment implies the development of a recirculation inside the room.

Based on Figures 9–11, the flow in all three investigated cases is parallel to the building envelope, due to the wind direction. In the case of the lowest wind speed, i.e., 6 m/s, the air flow is decelerated by the building's friction forces and the z-component of wind speed is very low; due to wind direction and the low wind speed. These are the main reasons why the case of 6 m/s wind speed presents the minimum mass flow rate among the three cases. By comparing the rest two cases (10 and 15 m/s wind speed), it can be concluded that the air flow with the maximum velocity is significantly deflected from the building envelope, since the thickness and the required length of fully-developed boundary layer increases, as the velocity increases too. Therefore, the values of the z-component of the velocity (towards the building's window) in the area of interest are higher in the case with 10 m/s wind speed compared to the one with 15 m/s. This results in higher mass flow rate in the case with 10 m/s wind speed compared to the case with 15 m/s.

Figure 12 presents the velocity magnitude and the direction of the airflow through the open area of the tilted window for the case of wind direction of 45° and all the wind speeds investigated. In all three cases, the main mass of air enters the room from the right and the top side of the opening, while it exits through the left one. Furthermore, the two cases with 10 and 15 m/s wind speed present almost identical vectors. However, a difference between these cases can be observed on the upper-left corner, where the flow from the room to the environment in the case of 10 m/s wind speed presents higher velocity compared to the case of 15 m/s, as a result of the increased mass flow rate and the moving of the natural plane of the flow towards the window top side.

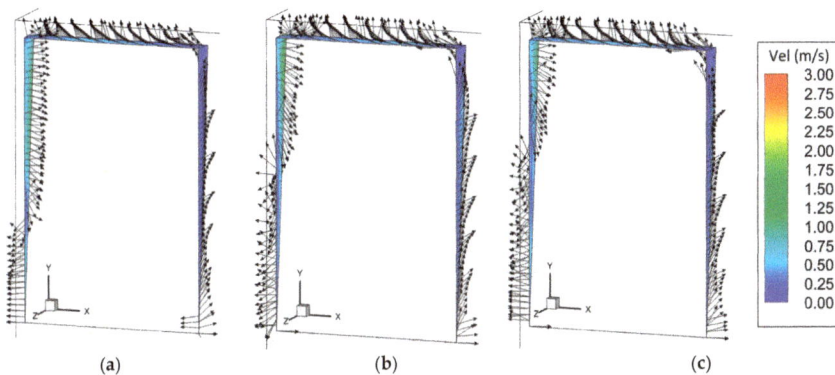

Figure 12. Vectors and velocity magnitude on the window opening for (**a**) 6 m/s, (**b**) 10 m/s, (**c**) 15 m/s wind speed, for 45° wind direction.

Figures 13–15 show the flow patterns and the horizontal profile of the z-component of velocity in the case of 215° wind direction at a height of 6 m along the y-axis for 6, 10 and 15 m/s wind speed, respectively.

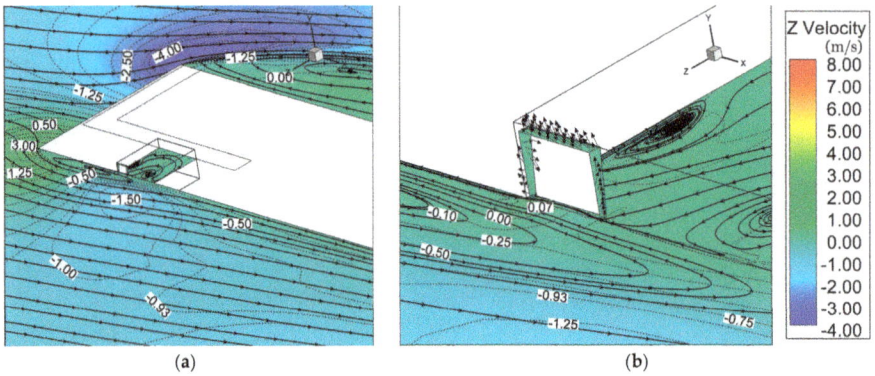

Figure 13. Flow patterns and contour of velocity (z-component) for 6 m/s wind speed and wind direction of 215°, (**a**) around the building, (**b**) in front of the tilted window.

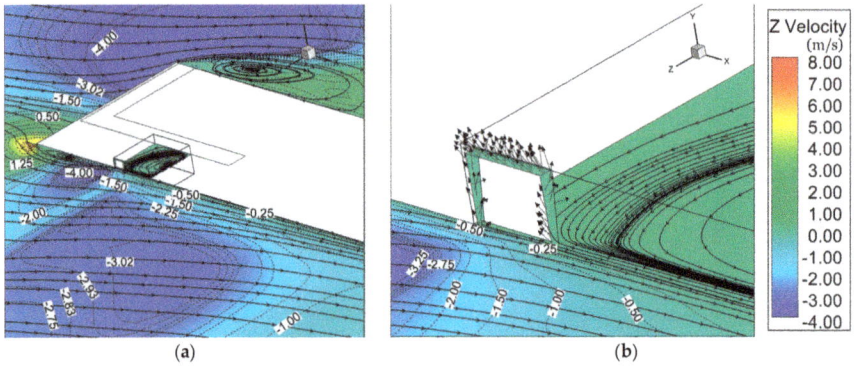

Figure 14. Flow patterns and contour of velocity (z-component) for 10 m/s wind speed and wind direction of 215°, (**a**) around the building, (**b**) in front of the tilted window.

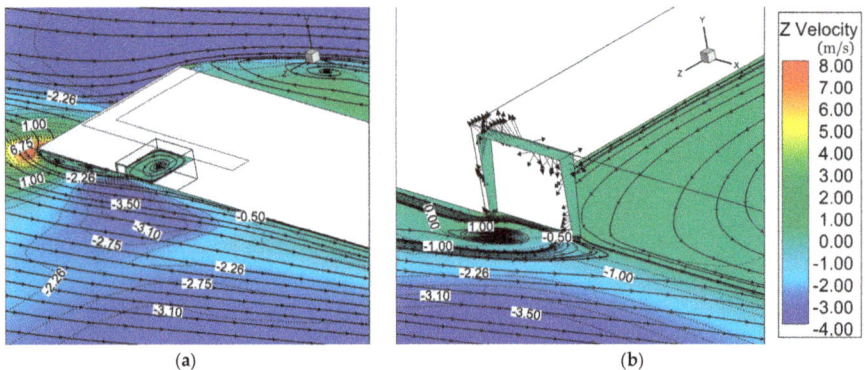

Figure 15. Flow patterns and contour of velocity (z-component) for 15 m/s wind speed and wind direction of 215°, (**a**) around the building, (**b**) in front of the tilted window.

As expected, a recirculation zone is developed close to the windward side of the building. This recirculation structure is transferred towards the window, as the wind speed increases. As already explained, when the wind speed increases, so does the thickness and the necessary length of fully-developed boundary layer. The development of a recirculation zone exactly in front of the window induces uncertainties in the developed flow field and significantly affects the induced mass flow rate through the respective opening. This is the main reason why the difference in the mass flow rate between the cases of 10 and 15 m/s wind speed is not so significant as in the other direction cases. Finally, the external wind direction affects the flow development inside the room, creating counterclockwise vortices.

Figure 16 presents the velocity magnitude and the direction of the airflow through the open area of the tilted window for the case of wind direction of 215° and all the wind speeds investigated. It can be observed that the recirculation developed in front of the window in the case of 15 m/s wind speed affects the developed flow field on the window opening, especially on the top side. Subsequently, in contrast to the rest two cases, where the flow clearly enters the room from the left and the top side of the opening, in the case of 15 m/s wind speed the flow on the top side is both inwards and outwards. This also affects the mass flow rate that enters the room (or equivalently exits the room) and the difference between the cases of 10 and 15 m/s air speeds is lower than the difference in other cases with different wind direction angles.

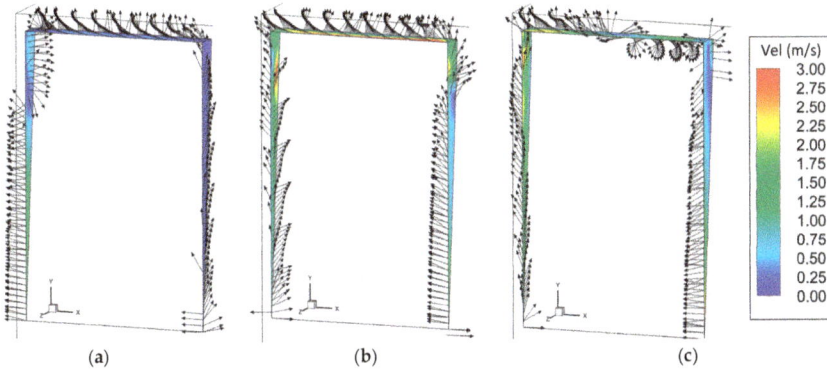

Figure 16. Vectors and velocity magnitude on the window opening for (**a**) 6 m/s, (**b**)10 m/s, (**c**)15 m/s wind speed, with a wind direction of 215°.

The fitted function depends on the type of window opening, building dimensions, and wind conditions. Thus, in the current study the best fit to the normalized data is achieved by using the Gaussian function provided by Equation (9):

$$f(x) = ae^{-\frac{(x-b)^2}{2c^2}} \tag{9}$$

The constants a and b are selected equal to 1 and 0, respectively, in order for the Gaussian curve to pass through the point (0,1). Depending on the incident velocity that characterizes each group of computational runs, the value of c parameter that ensures the best agreement between the derived dimensionless curves and the Gaussian one is c = 0.4055, c = 0.4292 and c = 0.3361 for the wind speeds of v = 6 m/s, v = 10 m/s and v = 15 m/s, respectively. The curve used for fitting for each specific velocity case along with the coefficient of determination, R^2, is given in Figure 17.

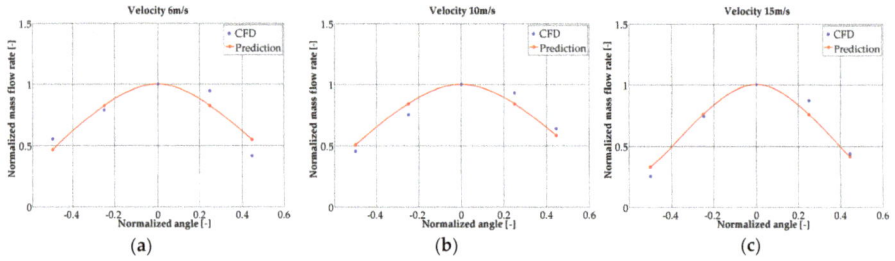

Figure 17. Fitting of data using Gaussian function: (**a**) case of 6 m/s wind speed, $R^2 = 0.8417$; (**b**) case of 10 m/s wind speed, $R^2 = 0.8892$; (**c**) case of 15 m/s wind speed, $R^2 = 0.9465$.

Because of the fact that the c parameter changes among the investigated cases, depending on the wind speed, it is necessary to formulate a mathematical expression that connects the c parameter with the velocity magnitude. This mathematical expression is given in Equation (10):

$$c(v) = -0.00273v^2 + 0.04956v + 0.20632 \tag{10}$$

where v is the wind speed in m/s.

The final expression of the mass flow rate prediction as a function of the normalized wind direction θ and wind speed v is described by the Equation (11):

$$f(\theta, v) = \exp\left(-\frac{\theta^2}{2(-0.00273v^2 + 0.04956v + 0.20632)^2}\right) \tag{11}$$

Finally, Equation (12) provides the maximum mass flow rate for each group of the examined cases as a function of the incident velocity, with $R^2 = 0.995$:

$$\dot{m}_{max}(v) = 0.04599v \tag{12}$$

The coefficient of determination of Equations (10) and (12) is equal to 1 and 0.995, respectively.

3.2. Verification of Empirical Function

In order to validate the derived mathematical correlation, it is necessary to conduct additional simulations using different wind speeds to assess the agreement of the provided values by the mathematical expression against the CFD results. In this framework, three values of velocity magnitude have been selected; the first one smaller than 6 m/s, the second one in-between the range of 6 and 15 m/s and the third one above the maximum selected value of 15 m/s.

Table 2 shows the percentage relative error between the mass flow rate numerically calculated and the one estimated by the derived function in the case of 2.5 m/s wind speed. The relative error has been calculated by Equation (13). The maximum numerical errors can be seen in the two limit values of wind direction, i.e., 45° and 215°, and are equal to 38.2% and 8.8%, respectively. Contrary to the limit values of wind speed, the interval ones present very good agreement. This fact can be also observed in Figure 18, which presents both the derived graph of the function for this specific wind speed (blue line) and the CFD results (red dots):

$$Relative\ error = \left|\frac{\dot{m}_{CFD} - \dot{m}_{prediction}}{\dot{m}_{CFD}}\right| \times 100\%, \tag{13}$$

Table 2. Percentage relative error between CFD and empirical normalized airflow rate for 2.5 m/s.

Angle (°)	Normalized Angle	Normalized Mass Flow Rate (CFD)	Prediction of Normalized Mass Flow Rate	Relative Error (%)
45°	−0.5	0.45	0.28	38.2
90°	−0.25	0.77	0.73	5.9
135°	0	1.00	1.00	0.0
180°	0.25	0.78	0.73	6.9
215°	0.45	0.33	0.36	8.8

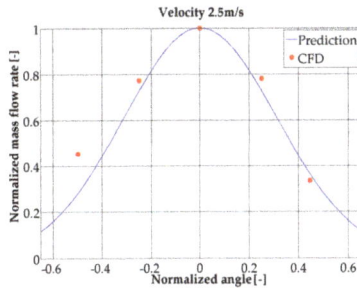

Figure 18. Comparison of normalized mass flow rate between CFD results and values derived by the function for 2.5 m/s wind speed.

The second wind speed that has been selected is equal to 12 m/s and is an interval value of the range between 6 and 15 m/s. It can be observed that in this case a better agreement between the CFD and empirical results can be achieved for all the wind directions, since the maximum relative difference is approximately equal to 3.5% (Table 3). This agreement is also noticeable in Figure 19, which presents both the derived graph of the function for this specific wind speed (blue line) and the CFD results (red dots).

Table 3. Percentage relative error between CFD and empirical normalized airflow rate for 12 m/s.

Angle (°)	Normalized Angle	Normalized Mass Flow Rate (CFD)	Prediction of Normalized Mass Flow Rate	Relative Error (%)
45°	−0.5	0.48	0.47	1.2
90°	−0.25	0.81	0.83	2.8
135°	0	1.00	1.00	0.0
180°	0.25	0.86	0.83	3.5
215°	0.45	0.54	0.55	1.2

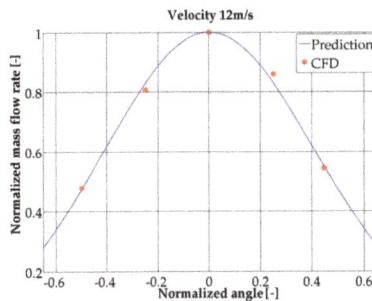

Figure 19. Comparison of normalized mass flow rate between CFD results and values derived by the function for 12 m/s wind speed.

For the last case, the wind speed is 16 m/s. In Table 4 the relative error between the estimated and CFD results can be seen. When the wind direction is parallel to the window i.e., 45° and 215° the relative error is significant, i.e., 35.5% and 36%, respectively. Moreover, an underestimation of mass flow rate can be observed for 90° and 180° as the velocity of 16 m/s does not belong to the range of selected values for Gaussian fitting. This fact can be also observed in Figure 20, which presents both the derived graph of the function for this specific wind speed (blue line) and the CFD results (red dots).

Table 4. Percentage relative error between CFD and empirical normalized airflow rate for 16 m/s.

Angle (°)	Normalized Angle	Normalized Mass Flow Rate (CFD)	Prediction of Normalized Mass Flow Rate	Relative Error (%)
45°	−0.5	0.39	0.25	35.5
90°	−0.25	0.77	0.71	8.3
135°	0	1.00	1.00	0.0
180°	0.25	0.77	0.71	8.1
215°	0.45	0.52	0.33	36.0

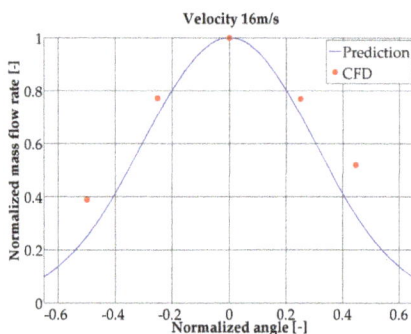

Figure 20. Comparison of normalized mass flow rate between CFD results and values derived by the function for 16 m/s wind speed.

Based on the results, it can be concluded that the mathematical correlation cannot accurately predict the mass flow rate through the window opening when both of the following two arguments are valid: a) the wind speeds are out of the range of the values that have been selected for the function development and b) the wind direction is parallel to the window's surface. This can be attributed to the fact that the Gaussian function is symmetrical around the central value, while the developed flow field presents recirculations and flow deflection in front of the window that induce uncertainties and dissimilarities, so the differences between the Gaussian function and the CFD results become more significant. Thus, it is difficult to predict the mass flow rate with reasonable accuracy in the cases of wind direction parallel to the window and wind speed out of the range of the selected values for the derivation of the function. These results are consistent with the findings of Wang [23], who observed that only when the mass flow is perpendicular to the tilted window the proposed semi-empirical model agreed with his CFD simulations. Furthermore, this model refers to opening angles 30° 45°, while in the current study an angle of 5.8° is examined. This is important because due to the complicated geometry it is not clear the location of the neutral plane which is a term in the semi-empirical model for the calculation of the ventilation rate. Thus, the results of this study cannot be compered by the semi-empirical model.

4. Conclusions

This paper presents a simple, versatile methodology for the development of an empirical equation, which can provide the air mass flow rate imposed by single-sided wind-driven ventilation of a room, as a function of external wind speed and direction, using the results from CFD simulations. k-w SST turbulence model and Boussinesq approximation have been used for the simulation of turbulence and buoyancy forces, respectively.

In order to achieve the derivation of a function from CFD simulations for prediction of the mass flow rate, it was necessary to use three wind speeds, namely 6, 10 and 15 m/s. In each case of wind speed five different wind directions were simulated. The normalized mass flow rates were fitted using a type of Gaussian function. The validation of the empirical function has been performed by conducting additional simulations with wind speed equal to 2.5, 12 and 16 m/s. In contrast to the case of the velocity of 12 m/s, whose predictions have a very good agreement with the simulation results, the other two cases present significant relative error when the airflow is parallel to the window. With these wind directions, the CFD results have showed the development of recirculations near the window and the deflection of the flow from the building. Since these phenomena are complicated and the function cannot accurately take them into account, the relative error between the simulation and the prediction in these cases is increased. Moreover, the selected velocities of 2.5 and 16 m/s do not belong in the range of the values which the correlation is based on. The broad range of wind speeds that have been examined and the non-symmetrical building formation (contrary to the symmetrical conditions of pilot simulations) distinguish the present work from previous publications, that use CFD simulations for an empirical correlation for a shorter range of wind speeds and symmetrical conditions. However, an interesting follow-up work of this study could be to use experimental data for further accuracy of the correlation, because in the present case the experimental values are not valid, since the weather station is located inside the developed boundary layer.

Author Contributions: In this work, the collaboration of many contributors were necessary. P.S. was responsible for the investigation, P.D. was responsible for the methodology, D.R. was responsible for the writing (review and editing) and N.N. was responsible for the project administration.

Funding: The analysis has been performed in the framework of PLUG-N-HARVEST research project, grant number 768735, funded by EU's Horizon2020 program.

Acknowledgments: This work has been carried out in the framework of the European Union's Horizon 2020 research and innovation program under grant agreement Plug'n'Harvest project. The authors would specially like to thank Verena Dannapfel and Tim Roeder from the University of Aachen for the fruitful collaboration.

Conflicts of Interest: The authors declare no conflict of interest.

Nomenclature

a	constant
A	constant, $N \cdot s^2 \cdot m^{-4}$
B	constant, $N \cdot s \cdot m^{-3}$
b	constant
D_ω	cross-diffusion term, $kg \cdot s^{-2} \cdot m^{-2}$
F	external body forces in momentum equation, $N \cdot m^{-3}$
G_ω	generation of ω, $kg \cdot s^{-2} \cdot m^{-2}$
$\tilde{G}_k$	generation of turbulence kinetic energy, $kg \cdot m^{-1} \cdot s^{-3}$
g_z	gravity vector, $9.81 m \cdot s^{-2}$
H	height, m
k	turbulent kinetic energy, $kg \cdot m^{-1} \cdot s^{-3}$
k_{eff}	effective conductivity, $W/(m \cdot K)$
L	length, m
$\dot{m}$	mass flow rate $kg \cdot s^{-1}$
$\hat{\dot{m}}$	normalized mass flow rate
p	pressure, Pa

S_k	user-defined source term, $kg \cdot m^{-1} \cdot s^{-3}$
S_ω	user-defined source term, $kg \cdot s^{-2} \cdot m^{-2}$
S_i	source term, $Pa \cdot m^{-1}$
T	temperature, K
t	time, s
v	Velocity vector, $m \cdot s^{-1}$
W	width, m
x_i	position vector with Cartesian components
Y_k	dissipation of **k**, $kg \cdot m^{-1} \cdot s^{-3}$
Y_ω	dissipation of ω, $kg \cdot s^{-2} \cdot m^{-2}$

Greek symbols

β	thermal expansion coefficient, K^{-1}
Γ_k	effective diffusivity of **k**, $kg \cdot m^{-1} \cdot s^{-1}$
Γ_ω	effective diffusivity ω, $kg \cdot m^{-1} \cdot s^{-1}$
θ	wind incident angle, °
$\hat{\theta}$	normalized wind incident angle
μ	dynamic viscosity, $kg \cdot m^{-1} \cdot s^{-1}$
ρ	density, $kg \cdot m^{-3}$
τ	stress tensor
ω	specific turbulent dissipation rate, $kg \cdot m^{-3} \cdot s^{-2}$

Subscripts

ref	reference

References

1. Allard, F.; Allard, F. *Natural Ventilation in Buildings: A Design Handbook*; James & James: London, UK, 1998.
2. American Society of Heating, Refrigerating and Air-Conditioning Engineers. *Ventilation and Infiltration: Chapter 26 in ASHRAE Fundamentals Handbook 2001.* Available online: https://www.thenbs.com/PublicationIndex/documents/details?Pub=ASHRAE&DocID=256357 (accessed on 25 April 2019).
3. Visagavel, K.; Srinivasan, P. Analysis of single side ventilated and cross ventilated rooms by varying the width of the window opening using CFD. *Sol. Energy* **2009**, *83*, 2–5. [CrossRef]
4. Etheridge, D. Design procedures for natural ventilation. In *Advanced Environmental Wind Engineering*; Springer: Berlin, Germany, 2016; pp. 1–24.
5. Passe, U.; Battaglia, F. *Designing Spaces for Natural Ventilation: An Architect's Guide*; Routledge: London, UK, 2015.
6. Linden, P.F. The fluid mechanics of natural ventilation. *Annu. Rev. Fluid Mech.* **1999**, *31*, 201–238. [CrossRef]
7. Van Hooff, T.; Blocken, B. Coupled urban wind flow and indoor natural ventilation modelling on a high-resolution grid: A case study for the Amsterdam ArenA stadium. *Environ. Model. Softw.* **2010**, *25*, 51–65. [CrossRef]
8. Qizhi, K.; Tan, H.; Zhu, W.; Li, M. Study on Influences of Wind Characteristics on Natural Ventilation Effect in Welding Plant. *Build. Energy Environ.* **2007**, *1*, 024. Available online: http://en.cnki.com.cn/Article_en/CJFDTOTAL-JZRK200701024.htm (accessed on 26 April 2019).
9. Aldawoud, A. Windows design for maximum cross-ventilation in buildings. *Adv. Build. Energy Res.* **2017**, *11*, 67–86. [CrossRef]
10. Hitchin, E.; Wilson, C. A review of experimental techniques for the investigation of natural ventilation in buildings. *Build. Sci.* **1967**, *2*, 59–82. [CrossRef]
11. Chen, Q. Ventilation performance prediction for buildings: A method overview and recent applications. *Build. Environ.* **2009**, *44*, 848–858. [CrossRef]
12. *The Ventilation of Buildings: Investigation of the Consequences of Opening One Window on the Internal Climate of a Room*; Report C 448; TNO Institute for Environmental Hygiene and Health Technology (IMG-TNO): Delft, The Netherlands, 1980; Available online: https://www.aivc.org/resource/ventilation-buildings-investigation-consequences-opening-one-window-internal-climate-room (accessed on 25 April 2019).

13. Warren, P. Ventilation through Openings on One Wall only, Energy Conservation in Heating, Cooling, and Ventilating Buildings. Heat and Mass Transfer Techniques and Alternatives. 1978. Available online: https://www.aivc.org/resource/ventilation-through-openings-one-wall-only (accessed on 26 April 2019).

14. Warren, P.; Parkins, L.M. *Single-Sided Ventilation through Open Window*. ASHRAE SP49. 1984. Available online: https://www.aivc.org/resource/single-sided-ventilation-through-open-windows (accessed on 25 April 2019).

15. American Society of Heating, Refrigerating and Air-Conditioning Engineers. *Ventilation and Infiltration: Chapter 27 in ASHRAE Fundamentals Handbook 2001*. 2005. Available online: https://www.globalspec.com/reference/53625/203279/chapter-27-ventilation-and-infiltration (accessed on 25 April 2019).

16. Allocca, C.; Chen, Q.; Glicksman, L.R. Design analysis of single-sided natural ventilation. *Energy Build.* **2003**, *35*, 785–795. [CrossRef]

17. Asfour, O.S.; Gadi, M.B. A comparison between CFD and Network models for predicting wind-driven ventilation in buildings. *Build. Environ.* **2007**, *42*, 4079–4085. [CrossRef]

18. Caciolo, M.; Stabat, P.; Marchio, D. Full scale experimental study of single-sided ventilation: analysis of stack and wind effects. *Energy Build.* **2011**, *43*, 1765–1773. [CrossRef]

19. Larsen, T.S.; Heiselberg, P. Single-sided natural ventilation driven by wind pressure and temperature difference. *Energy Build.* **2008**, *40*, 1031–1040. [CrossRef]

20. Dascalaki, E.; Santamouris, M.; Argiriou, A.; Helmis, C.; Asimakopoulos, D.N.; Papadopoulos, K.; Soilemes, A. On the combination of air velocity and flow measurements in single sided natural ventilation configurations. *Energy Build.* **1996**, *24*, 155–165. [CrossRef]

21. Caciolo, M.; Cui, S.; Stabat, P.; Marchio, D. Development of a new correlation for single-sided natural ventilation adapted to leeward conditions. *Energy Build.* **2013**, *60*, 372–382. [CrossRef]

22. Tang, Y.; Li, X.; Zhu, W.; Cheng, P. Predicting single-sided airflow rates based on primary school experimental study. *Build. Environ.* **2016**, *98*, 71–79. [CrossRef]

23. Wang, H.; Chen, Q. A new empirical model for predicting single-sided, wind-driven natural ventilation in buildings. *Energy Build.* **2012**, *54*, 386–394. [CrossRef]

24. Wang, H.; Karava, P.; Chen, Q. Development of simple semiempirical models for calculating airflow through hopper, awning, and casement windows for single-sided natural ventilation. *Energy Build.* **2015**, *96*, 373–384. [CrossRef]

25. Pan, W.; Liu, S.; Li, S.; Cheng, X.; Zhang, H.; Long, Z.; Zhang, T.; Chen, Q. A model for calculating single-sided natural ventilation rate in an urban residential apartment. *Build. Environ.* **2019**, *147*, 372–381. [CrossRef]

26. Nikolopoulos, N.; Nikolopoulos, A.; Larsen, T.S.; Nikas, K.-S.P. Experimental and numerical investigation of the tracer gas methodology in the case of a naturally cross-ventilated building. *Build. Environ.* **2012**, *56*, 379–388.

27. Nikas, K.-S.; Nikolopoulos, N.; Nikolopoulos, A. Numerical study of a naturally cross-ventilated building. *Energy Build.* **2010**, *42*, 422–434. [CrossRef]

28. Larsen, T.S.; Nikolopoulos, N.; Nikolopoulos, A.; Strotos, G.; Nikas, K.-S. Characterization and prediction of the volume flow rate aerating a cross ventilated building by means of experimental techniques and numerical approaches. *Energy Build.* **2011**, *43*, 1371–1381. [CrossRef]

29. Boltzmann Number. Available online: http://www.thermopedia.com/content/207/ (accessed on 25 April 2019).

30. Patankar, S. *Numerical Heat Transfer and Fluid Flow*; CRC press: Boca Raton, FL, USA, 1980.

31. SIMPLE Algorithm. Available online: https://www.sharcnet.ca/Software/Ansys/17.0/en-us/help/flu_th/flu_th_sec_uns_solve_pv.html (accessed on 26 April 2019).

32. Launder, B.E.; Spalding, D.B. The numerical computation of turbulent flows. In *Numerical Prediction of Flow, Heat traNsfer, Turbulence and Combustion*; Elsevier: Amsterdam, The Netherlands, 1983; pp. 96–116.

33. van Hooff, T.; Blocken, B.; Tominaga, Y. On the accuracy of CFD simulations of cross-ventilation flows for a generic isolated building: comparison of RANS, LES and experiments. *Build. Environ.* **2017**, *114*, 148–165. [CrossRef]

34. Ramponi, R.; Blocken, B. CFD simulation of cross-ventilation for a generic isolated building: impact of computational parameters. *Build. Environ.* **2012**, *53*, 3448. [CrossRef]

35. Menter, F.R. Two-equation eddy-viscosity turbulence models for engineering applications. *AIAA J.* **1994**, *32*, 1598–1605. [CrossRef]

36. Kays, W.M.; Crawford, M.E.; Weigand, B. *Convective Heat and Mass Transfer*; McGraw-Hill Science: New York, NY, USA, 2005.
37. Franke, J. *Best Practice Guideline for the CFD Simulation of Flows in the Urban Environment*; Meteorological Inst, 2007. Available online: http://theairshed.com/pdf/COST%20732%20Best%20Practice%20Guideline%20May%202007.pdf (accessed on 25 April 2019).
38. Cedar Lake Ventures. Available online: https://weatherspark.com/y/54659/Average-Weather-in-Aachen-Germany-Year-Round (accessed on 25 April 2019).
39. Climate Aachen. Available online: https://www.meteoblue.com/en/weather/forecast/modelclimate/aachen_germany_3247449 (accessed on 25 April 2019).

energies

MDPI

Article

Effects of the Heat Treatment in the Properties of Fibrous Aerogel Thermal Insulation

Ákos Lakatos [1,*], Attila Csík [2], Anton Trník [3,4] and István Budai [5]

[1] Department of Building Services and Building Engineering, Faculty of Engineering, University of Debrecen, Ótemető str 2-4 1, 4028 Debrecen, Hungary

[2] Institute for Nuclear Research, Hungarian Academy of Sciences, Bem tér 18/c, 4026 Debrecen, Hungary; csik.attila@atomki.mta.hu

[3] Department of Physics, Faculty of Natural Sciences, Constantine the Philosopher University in Nitra, Tr. A Hlinku 1, 94974 Nitra, Slovakia; atrnik@ukf.sk

[4] Department of Materials Engineering and Chemistry, Faculty of Civil Engineering, Czech Technical University in Prague, Thákurova 7, 16629 Prague, Czech Republic

[5] Department of Engineering Management and Enterprise, Faculty of Engineering, University of Debrecen, Ótemető str 2-4, 4028 Debrecen, Hungary; budai.istvan@eng.unideb.hu

* Correspondence: alakatos@eng.unideb.hu; Tel.: +36-3033-468-61

Received: 29 April 2019; Accepted: 22 May 2019; Published: 25 May 2019

Abstract: Nowadays, besides the use of conventional insulations (plastic foams and wool materials), aerogels are one of the most promising thermal insulation materials. As one of the lightest solid materials available today, aerogels are manufactured through the combination of a polymer with a solvent, forming a gel. For buildings, the fiber-reinforced types are mainly used. In this paper, the changes both in the thermal performance and the material structure of the aerogel blanket are followed after thermal annealing. The samples are put under isothermal heat treatments at 70 °C for weeks, as well as at higher temperatures (up to 210 °C) for one day. The changes in the sorption properties that result from the annealing are presented. Furthermore, the changes in the thermal conductivity are followed by a Holometrix Lambda heat flow meter. The changes in the structure and surface of the material due to the heat treatment are investigated by X-ray diffraction and with scanning electron microscopy. Besides, the above-mentioned measurement results of differential scanning calorimetry experiments are also presented. As a result of using equipment from different laboratories that support each other, we found that the samples go through structural changes after undergoing thermal annealing. We manifested that the aerogel granules separate down from the glass fibers and grow up. This phenomenon might be responsible for the change in the thermal conductivity of the samples.

Keywords: fibrous aerogel; thermal conductivity; heat treatment; XRD; SEM; DSC

1. Introduction

A method to decrease the energy use as well as the emission of greenhouse gases is to apply thermal insulation materials. [1–4] Silica aerogels are mesoporous materials that have very small thermal conductivity (λ, W/mK) [5]. Aerogel materials are applied as thermal excellent insulations. It has been presented in recent papers that fibrous, transparent, or translucent aerogel-type insulations are some of the most innovative/advanced insulation materials. Aerogel is a porous material with low density, and cells on the nanoscale. It is manufactured from different types of gels by supercritical drying; silica alco-gels are the most widely manufactured, which are most regularly prepared from silica gel [6–11]. In many papers, in the systematic literature, investigations of aerogel-related contents can be found, but due to a lack of space, only some of them can be presented as well as processed in the introduction part of this paper [12–14]. The main goal of this paper is to present and understand experimental results carried out on glass fiber-reinforced aerogel samples. The thermo-hygric properties of the aerogel have been comprehensively presented in recent papers [15–20].

It was previously presented in [6] that through service time, the thermal conductivity of the samples varies. On the nanometric and microscopic scale, it strongly depends on the distribution and size of the fibers.

It should be stated that over a temperature interval (−15 °C to 25 °C), the effect of air temperature has a negligible or only minor influence on the thermal conductivity coefficient of a sample; nevertheless, at uncommon temperature varieties (25 °C < T < 200 °C), significant changes are expected. Near these mentioned temperatures, which can be thought to be thermal, annealing, materials can suffer physical or chemical changes and can also provoke a rise in thermal conductivity.

High temperature can raise the kinetic reaction rates inside the materials, and can probably boost the chemical, as well as the physical degradation phenomenon [21]. In industrial cases, higher (elevated) temperature impacts can reach the materials. By insulating pipes, transporting hot fluid or steam, or heating, ventilation, and air conditioning (HVAC) systems with temperatures of up to 200 °C–300 °C can be formed.

The stable thermal resistance is a property that is required for formulating the design insulation performance under service conditions. Controlled aging is a method for modeling the chemical, mechanical, and thermal (physical) properties of a sample as a function of time. Whether thermal annealing causes elevated aging regarding the silica aerogel bats is unknown, so measurements and models are necessary. Having a relatively high melting point, general temperature ranges (−10 °C to 60 °C) would not have a substantial influence, but the higher temperatures would have. Jelle et al., in their paper, presented thermal exposures as aging issues, for example: the thermal annealing [21,22]. Miros revealed that heat annealing as a thermal aging procedure influenced the thermal properties of mineral wool materials at unconventional temperature ranges (100 °C to 600 °C) [23]. Furthermore, for insulation such as rock wools or fibrous silica aerogels, the deterioration of the thermal performance should appear through chemical or physical reactions as structural modifications. Similar to the paper of Miros [23], where aging caused structural changes followed by variations in the thermal conductivity. They explored changes with a heat flow meter.

In recent years, besides the measurements of the thermal conductivities and sorption isotherms under normal (general conditions), test results on heat-treated samples were presented as well. These ordinary measurements are usually completed with destructive or non-destructive structural investigations e.g., scanning electron microscopy, calorimetry, and diffraction for crystallization study. It has been well presented in previous papers [24,25] that scanning electron microscopy (SEM) combined with element mapping could illustrate essential information regarding the structures and orders of the fibers [26,27]. After annealing, an analysis of the microstructure of the samples is very important. Moreover, besides these investigations, tests with an X-ray diffractometer (XRD) should be carried out. Several papers [28,29] present the XRD measurements of the aerogels after annealing at elevated temperatures, which are usually over 500 °C. Also, differential scanning calorimetry (DSC) investigations combined with scanning electron microscopy as well as with sorption measurements were executed on aerogels [30,31].

In this study, we systematically investigated the influence of different aging and temperatures on the structure and thermal conductivity of glass fiber-enhanced silica aerogel, with 150 kg/m^3 density. Then, changes in the sorption isotherms as well as modification in the structure and in the thermal properties are introduced after executing SEM, DSC, and XRD tests. The samples applied in this study were received from a Hungarian distributor.

2. Materials and Methods

2.1. Thermal Conductivity Measurements

The measurement of the thermal conductivity of insulation materials can be performed by following the rules of the EN ISO 12664:2001 standard [32] (Thermal performance of building materials and products. Determination of thermal resistance by means of guarded hot plate and heat flow meter methods. Dry and moist products of medium and low thermal resistance) [24,29]. In order to reveal the exact thermal conductivity coefficient of our samples, a Holometrix Lambda 2000 type heat flow meter (HFM) (Bedford, Massachusetts, US) was applied. The used equipment was manufactured to specify the thermal conductivity coefficient of insulation materials with about 5% accuracy, which is in agreement with standards ASTM C518 [33] and ISO 8301 [34]. As the basis of the research, the thermal conductivity measurement results were presented in a previous paper [6] and here, these results will be improved through a deeper engineering analysis of the materials. The measurement conditions can be found in [6,18,35–37].

2.2. Sorption Isotherm Measurements

In order to find the moisture/vapor sorption properties of the samples, three pieces of equipment should be combined: a drying apparatus, a climatic chamber, and a milligram precision balancer. First, in order to find the sorped amount of water, the materials should be desiccated to a changeless weight, which in our case is a Venticell 111 apparatus (MMM Medcenter Facilities GmbH, Münich, Germany). It works with hot air between 10–250 °C. By using the equipment, the material can be dried; furthermore, samples can be thermally (heat) treated by setting different temperature values [6,18,35–37].

After treating the samples in this heater/dryer to constant mass, their weights must be registered with a balance. Afterwards, samples need to be positioned in a climatic chamber, which in our case is a Climacell 111 apparatus (MMM Medcenter Facilities GmbH, Münich, Germany). The apparatus can fix the temperature between 0–100 °C with any humidity between 10–90%.

For recording the sorption isotherms curves, four aerogel samples with 1-cm thickness and a 10 cm × 10 cm base area were tested. The investigations were done by following the rules of ISO 12571: 2013 standard [38] (Hygrothermal performance of building materials and products—Determination of hygroscopic sorption properties, Part B—climatic chamber method). In the humidity chamber, the temperature was fixed to 23 ± 1 °C, while the relative humidities were 30%, 50%, 65%, 80%, and 90%. From the wet and dry masses of all the samples, the moisture contents were calculated, and the results were averaged. The sorption isotherm of the unannealed sample was referred to in our previous work [6], while the graph for the two annealed samples is presented here.

2.3. Differential Scanning Calorimetry Measurements

The specific heat capacity (c_p) was determined according to standard DIN 51007 [39], where a differential scanning calorimetry (DSC) and a sapphire as the calibrant are used. DSC was carried out using a DSC 822e (Mettler Toledo, Greifensee, Switzerland) device. The experiments were done in the temperature interval from 25 °C up to 300 °C in an air atmosphere with a flow rate of 50 mL/min. The heating rate was 10 °C/min and the isothermal regime (5 min) was applied before and after linear heating. The mass of samples was about 9 mg, and aluminum crucibles (volume of 40 mm^3) with lids were used.

The specific heat capacity was calculated from the measured heat flow data of empty crucibles, a sapphire, and a sample using the equation:

$$c_p = c_{pc}(m_c(A_s - A_b))/(m_s(A_c - A_b)) \qquad (1)$$

where A_b, A_s, and A_c (in W/g) are the measured amplitudes (heat flows) of the empty crucibles, sample, and sapphire, respectively. Moreover, mc and m_s (in g) are the masses of the sapphire and sample, respectively, and c_{pc} (in J/(kg·K)) is the specific heat capacity of the sapphire [34].

2.4. Heat Treatments of the Samples

The heat-treating of the samples was executed in the VentiCell 111 apparatus (MMM Medcenter Facilities GmbH, München, Germany), in which the samples can be heat treated or dried at different air temperatures ranging between 10–250 °C. The equipment works with air circulation using an inbuilt ventilator [6,18,35–37]. The measurements were performed according to the ISO 12664 [32] and 12667 [40] standards. The tests were executed before and after thermal annealing the samples. Two different thermal annealing rows were done. The samples were put under isothermal heat treatments at 70 °C for six weeks (isothermal heat treatment); they were also put under thermal treatment at higher temperatures (70 °C, 100 °C, 130 °C, 150 °C, 180 °C, and 210 °C) for one day on the same sample, one after the other.

2.5. X-ray Diffraction Experiments

Crystallographic information from the as-received and annealed samples has been obtained by applying the X-ray diffraction (XRD) method. The measurements were performed by a Siemens diffractometer (Kristalloflex 710H, Siemens, Karlsruhe, Germany) using CuKα irradiation with $\lambda = 0.154$ nm. Scanning geometry between θ–2θ was applied to perform the measurements; the X-ray tube was operated with 40 mA and 40 kV settings. The high-angle spectra were measured between 20–85° to study the presence of solid phases in the samples.

2.6. Scanning Electron Microscopy Investigations

The materials were analyzed using a Hitachi TM-3030 scanning electron microscope (SEM) (Hitachi High-Technologies Europe GmbH, Japan) with a Bruker Quantax 70 EDS system (Berlin, Germany). The TM3030 is equipped with premium signal detectors that have been incorporated in field emission (FE-SEM) and variable-pressure (VP-SEM) and provide unparalleled image quality. The detectors can be effectively operated under low-vacuum conditions and can support high-sensitivity four-segment semiconductor (BSE) detector for image observation without metal coating. The TM-3030 can be used to observe BSE images with a magnification range from 15 to 30,000×. Element mapping (energy-dispersive spectroscopy, EDS) with X-ray analysis with a wide detection area (30 mm²) was executed on the samples as well.

3. Results

3.1. Thermal Conductivity Results

In a recent paper [6], it was presented that the thermal conductivity remains constant after annealing the samples until 42 days under atmospheric pressure and at 70 °C. It was found that the thermal conductivity ranges around 0.017 W/(m·K). This thermal conductivity value was equal to that measured on the as-received (unannealed) sample. However, after annealing the samples at 100 °C, 150 °C, 180 °C, and 210 °C for one day (one after the other on the same sample), the thermal conductivity was changed. Between 100–200 °C, it changes up to 0.018 W/(m·K), while above 200 °C, it jumps up to 0.020 W/(m·K) [6]. In order to go deeper into the reasons of the changes, further experiments were executed.

3.2. Results of Sorption Isotherm Measurements

Sorption isotherm investigations were executed with the above-mentioned method. In Figure 1, the sorption isotherm curves (equilibrium moisture content versus relative humidity at 23 °C) of annealed samples at 70 °C for 42 days and the graph of the thermally treated sample at 100 °C, 150 °C, 180 °C, and 210 °C for one day are visible. The sorption isotherm of the unannealed samples was also presented in [6]. The shape of the sorption isotherm of the as-received sample shows continuously increasing moisture content with the increasing relative humidity, and without any breaks in it. This shape presents the type II isotherm of the new classification of the BET isotherm representing adsorption isotherms on macroporous adsorbents with strong affinities [41].

Interestingly, one can see that the shapes of the sorption isotherms of the annealed samples are different. Between 50–75% relative humidity, the isotherms have a strong break. After analyzing the shape of the sorption isotherms, one can state that these shapes can be identified as type-IV isotherms, which characterizes mesoporous adsorbents with strong affinities [41]. We should conclude here that some structural change occurred due to the thermal annealing.

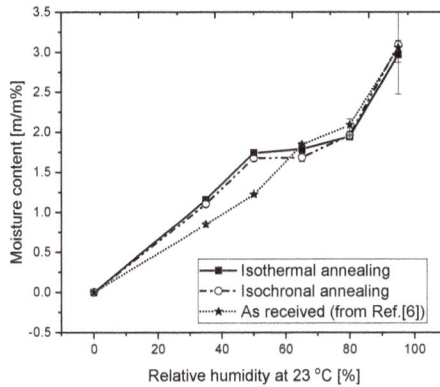

Figure 1. The sorption isotherm graphs.

3.3. Scanning Electron Microscope Results

To visualize the expected changes in the structures, scanning electron microscope measurements were executed. Pieces of unannealed, isothermally annealed, and isochronally annealed samples were put under an SEM test into the microscope. The results are presented in Figures 2–5 as: (Figure 2) an SEM image of the as-received sample, (Figure 3) an EDS element map of the as-received sample, (Figure 4) an SEM image of the isothermally annealed sample, and (Figure 5) an SEM image of the isochronally annealed sample. From the SEM pictures (Figure 2), it can be seen that the aerogel granules are joined to the fibers, which is in good connection with the other images presented in the literature [24,26].

Figure 2. SEM image of the as-received sample.

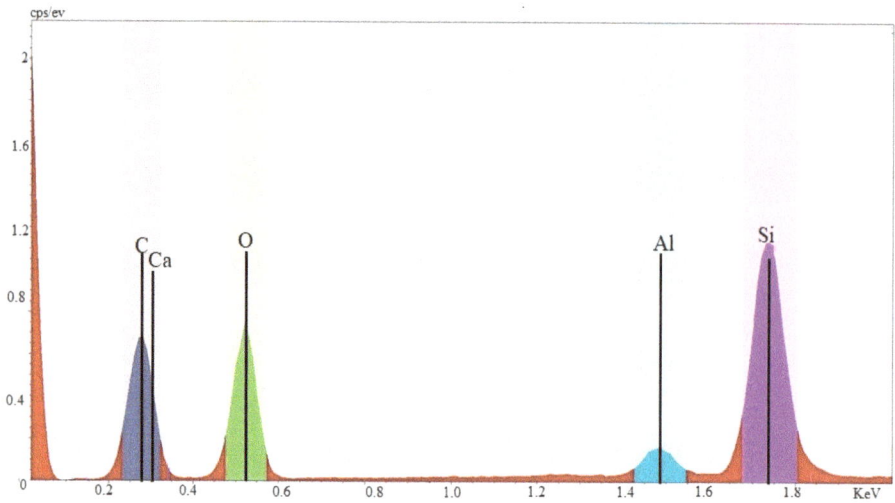

Figure 3. Energy-dispersive spectroscopy (EDS) element map of the as-received sample.

From the element maps, one can observe that the main components of the sample—silicon, oxygen, aluminum, and some calcium and carbon contaminants—may be coming from the glass fibers.

From the obtained SEM images, we can conclude that the particles after annealing are separated from the fibers and create bigger granulates. It can be clearly stated that the structure of the samples is changing due to the heat treatment. After thermal annealing at high temperatures, one cannot observe any particles on the fibers, while aerogel particles are visible on the surface of the fibers of the as-received and isothermally annealed samples. A similar effect was predicted by Li et al. and Yang et al. [24,26]. Iswar et al. visualized similar effects with SEM after annealing the samples at 65 °C from 2 h to 24 h [5]. Liang et al. [42] postulated that the investigations of the service life of aerogel-based insulations can be achieved by scanning electron microscope, too.

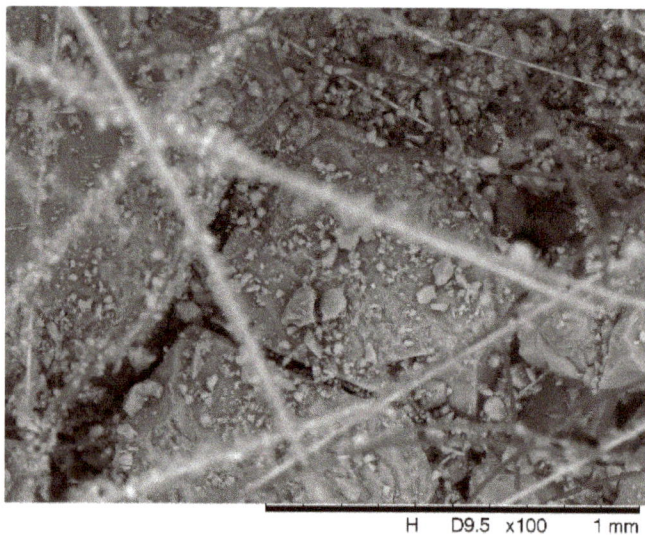

Figure 4. SEM image of the isothermally annealed sample.

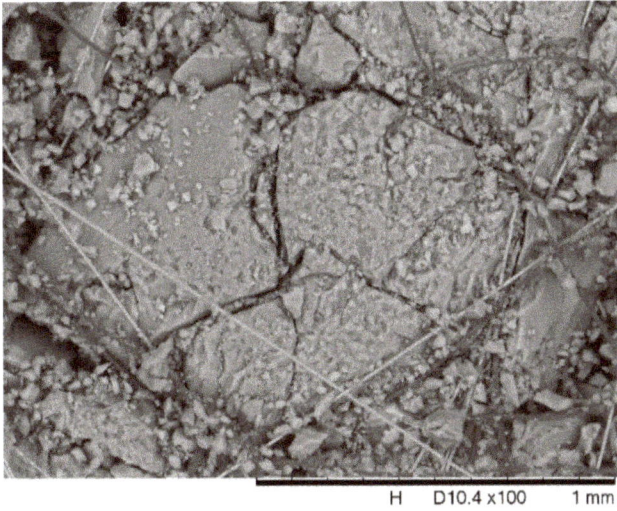

H   D10.4 x100   1 mm

Figure 5. SEM image of the isochronally annealed sample.

3.4. Differential Scanning Calorimetry Results

Figure 6 shows the DSC results of an as-received sample.

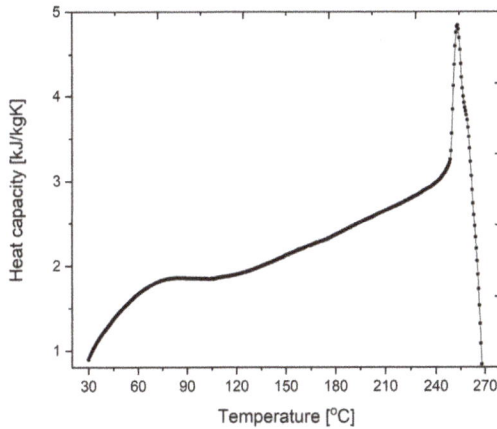

Figure 6. The differential scanning calorimetry (DSC) results of the aerogel sample.

From Figure 6, it is visible that the specific heat capacity at room temperature is about 1000 J/(kg·K), while it is constant almost between 70–120 °C. After this, a continuously increasing part is visible, while a great change is observable above 240 °C. Here, some reaction might take place. Hasan et al. [30] presented combined measurements on aerogel samples, including calorimetry tests. They showed a critical region between 130–260 °C.

3.5. Results of XRD Measurements

In order to further investigate our samples and check structural changes, XRD measurements were performed on all three types of samples. As it was already presented in [28,29,43,44], XRD is reasonable tool to understand the structural form of the aerogel samples. The results are presented in

Figures 7–9 as: (Figure 7) XRD spectra of the as-received sample, (Figure 8) the XRD spectra of the isothermally annealed samples, and (Figure 9) the XRD spectra of the isochronally annealed samples. Zhu et al., Kwon and Choi, and Music et al. [29,43,44] show the characteristic peak near 22–23° that belong to the amorphous SiO_2 phase. Additionally, Zhu et al. [28] further observed a small peak near 43° as it is also visible in our results (Figure 7), where the XRD spectra of the as-received sample are presented.

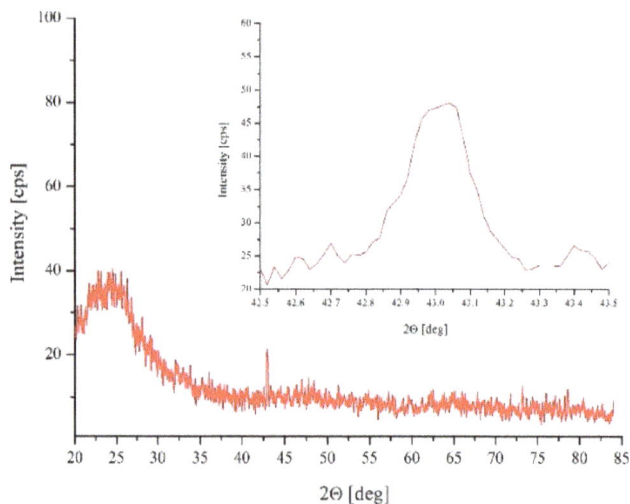

Figure 7. XRD spectra of the as-received sample.

From Figure 7, one can state that a broad peak can be found at about 22°, which belongs to the amorphous SiO_2, while a smaller peak is observable at 43° that relates to the carbon content. These results are in good agreement with the results presented in [28]. The carbon content can originate as the side effect of the preparation (super critical CO_2 extraction). In Figures 8 and 9, the XRD results of the annealed samples can be found.

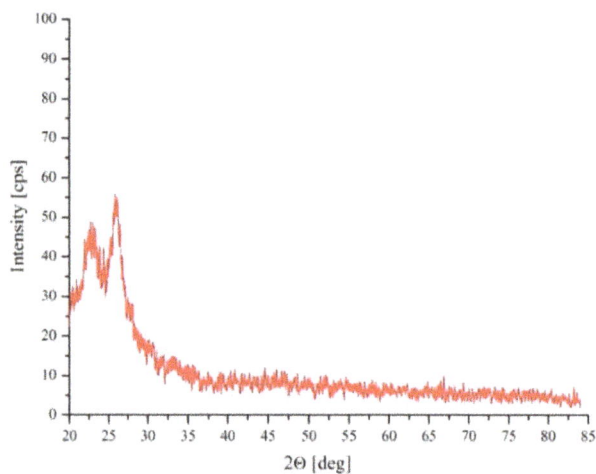

Figure 8. The X-ray diffractometer (XRD) spectra of the isothermally annealed samples.

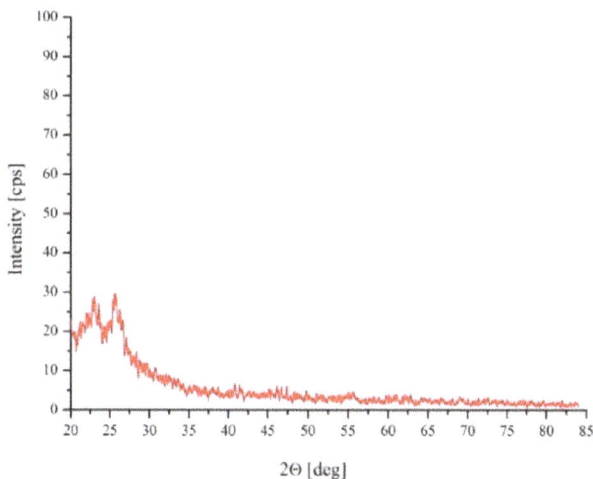

Figure 9. The X-ray diffractometer (XRD) spectra of the isochronally annealed samples.

From both graphs (Figures 8 and 9), one can see that the peak at 43° disappears as a result of annealing, while besides the broad peak observed on as-prepared sample near 22°, a new peak appears. These twin peaks were reported by Zhou et al. [28]. From these results, we could expect that heat treatment in such materials causes either a crystallization process or the growth of the grains. In Reference [45], it was stated that the different types of measurements carried out on fiber aerogel samples are very important, as they support each other. Reference [46] stated that the artificial aging of insulation materials and their investigations with different methods are very important.

4. Conclusions

Silica aerogels present acceptable thermal conductivity, and they are mainly used as thermal superinsulators. Here, we studied the influence of aging and drying processes on the microstructure and thermal properties of aerogel samples. Glass wool–silica gel composites were aged (thermal annealed) for variable temperatures and times at ambient air. Different experimental results were presented, supporting each other, and were carried out on glass fiber-reinforced aerogel samples. The tests of different laboratories supported the main conclusion of this paper. We can declare that thermal aging has an effect on the properties of the insulation materials. We observed the microstructural changes by sorption measurements through the variations of the shapes of the curves as well as by scanning electron microscopy; then, we manifested the growing of the grains. Furthermore, with X-ray diffraction, and last but not least with differential scanning calorimetry, we presented the changes in the phases and the presence of kinetic reactions. We pointed out that the aerogel grains are growing due to the thermal annealing and separating from the glass fibers. These results were combined with previously published ones, where the changes in the thermal conductivity caused by the thermal annealing were manifested. These results could provide a good base for the designers to know the applicability limit of the samples in high-temperature places.

Author Contributions: Conceptualization, Á.L.; Investigation, Á.L, A.C, I.B. and A.T.; Methodology, Á.L.; Supervision, Á.L.; Writing—original draft, Á.L., A.C., and A.T.; Writing-review & editing, Á.L., A.C. and A.T.

Funding: The research was financed the Ministry of Human Capacities in Hungary (grant number 20428-3/2018/FEKUTSTRAT).

Acknowledgments: The research was financed by the Higher Education Institutional Excellence Programme of the Ministry of Human Capacities in Hungary, within the framework of the Energetics thematic programme of the University of Debrecen. The author A.T. gratefully acknowledges a financial support for his work from RVO: 11000.

Conflicts of Interest: The authors declare no conflict of interest.

References

1. Kalmár, F. Energy analysis of building thermal insulation. In Proceedings of the 11th Conference for Building Physics, Dresden, Germany, 26–30 September 2002; pp. 103–112.
2. Kunic, R. Carbon footprint of thermal insulation materials in building envelopes. *Energy Effic.* **2017**, *10*, 1511–1528.
3. Szabó, G.L.; Kalmár, F. Parametric analysis of buildings' heat load depending on glazing—Hungarian case study. *Energies* **2018**, *11*, 3291. [CrossRef]
4. Csáky, I.; Kalmár, F. Investigation of the relationship between the allowable transparent area, thermal mass and air change rate in buildings. *J. Build. Eng.* **2017**, *12*, 1–7. [CrossRef]
5. Iswara, S.; Griffa, M.; Kaufmann, R.; Beltrand, M.; Hubera, L.; Brunner, S.; Lattuad, M.; Koebel, M.M.; Malfait, W.J. Effect of aging on thermal conductivity of fiber-reinforced aerogel composites: An X-ray tomography study. *Microporous Mesoporous Mater.* **2019**, *278*, 289–296. [CrossRef]
6. Lakatos, Á. Stability investigations of the thermal insulating performance of aerogel blanket. *Energy Build.* **2019**, *139*, 506–516. [CrossRef]
7. Aegerter, A.; Leventis, N.; Koebel, M. *Aerogels Handbook*; Springer: New York, NY, USA, 2011; pp. 537–564.
8. Hostler, S.R.; Abramson, A.R.; Gawryla, M.D.; Bandi, S.A.; Schiraldi, D.A. Thermal conductivity of a clay-based aerogel. *Int. J. Heat Mass Transf.* **2008**, *52*, 665–669. [CrossRef]
9. Schultz, J.M.; Jensen, K.I.; Kristiansen, F.H. Super insulating aerogel glazing. *Solar Energy Mater. Solar Cells* **2005**, *89*, 275–285. [CrossRef]
10. Galliano, R.; Wakili, K.G.; Stahl, T.; Binder, B.; Daniotti, B. Performance evaluation of aerogel-based and perlite-based prototyped insulations for internal thermal retrofitting: HMT model validation by monitoring at demo scale. *Energy Build.* **2016**, *12*, 275–286. [CrossRef]
11. Wakili, K.G.; Stahl, T.; Heiduk, E.; Schuss, M.; Vonbank, R.; Pont, U.; Sustr, C.; Wolosiuk, D.; Mahdavi, D. High performance aerogel containing plaster for historic buildings with structured façades. *Energy Procedia* **2015**, *78*, 949–954. [CrossRef]
12. Lucchi, E.; Becherini, F.; Di Tuccio, M.C.; Troi, A.; Frick, J.; Roberti, F.; Hermann, C.; Fairnington, I.; Mezzasalma, G.; Pockelé, L.; et al. Thermal performance evaluation and comfort assessment of advanced aerogel as blown-in insulation for historic buildings. *Build. Environ.* **2017**, *122*, 258–268. [CrossRef]
13. Lucchi, E.; Roberti, F.; Alexandra, T. Definition of an for the thermal performance evaluation of inhomogeneous walls. *Energy Build.* **2018**, *179*, 99–111. [CrossRef]
14. Berardi, U.; Lakatos, Á. Thermal bridges of metal fasteners for aerogel-enhanced blankets. *Energy Build.* **2019**, *185*, 307–315. [CrossRef]
15. Stahl, T.; Brunner, S.; Zimmermann, M.; Wakili, K.G. Thermo-hygric properties of a newly developed aerogel based insulation rendering for both exterior and interior applications. *Energy Build.* **2012**, *44*, 114–117. [CrossRef]
16. Cuce, E.; Cuce, P.M.; Wood, C.J.; Riffat, S.B. Toward aerogel based thermal superinsulation in buildings: A comprehensive review. *Renew. Sustain. Energy Rev.* **2014**, *34*, 273–299. [CrossRef]
17. Baetens, R.; Jelle, B.P.; Gustavsen, A. Aerogel insulation for building applications: A state-of-the-art review. *Energy Build.* **2011**, *43*, 761–769. [CrossRef]
18. Lakatos, Á. Investigation of the moisture induced degradation of the thermal properties of aerogel blankets: Measurements, calculations, simulations. *Energy Build.* **2017**, *139*, 506–516. [CrossRef]
19. Hoseini, A.; Bahrami, M. Effects of humidity on thermal performance of aerogel insulation blankets. *J. Build. Eng.* **2017**, *13*, 107–115. [CrossRef]
20. Nosrati, R.H.; Berardi, U. Hygrothermal characteristics of aerogel-enhanced insulating materials under different humidity and temperature conditions. *Energy Build.* **2018**, *158*, 698–711. [CrossRef]
21. Jelle, B.P. Accelerated climate ageing of building materials, components and structures in the laboratory. *J. Mater. Sci.* **2012**, *47*, 6475–6496. [CrossRef]

22. Jelle, B.P. Traditional, state-of-the-art and future thermal building insulation materials and solutions—properties, requirements and possibilities. *Energy Build.* **2011**, *43*, 2549–2563. [CrossRef]

23. Miros, A. Thermal Aging Effect on Thermal Conductivity Properties of Mineral Wool Pipe Samples at High Temperature. In Proceedings of the 3rd World Congress on Mechanical, Chemical, and Material Engineering (MCM'17), Rome, Italy, 8–10 June 2017.

24. Yang, X.; Sun, Y.; Shi, D.; Liu, J. Experimental investigation on mechanical properties of a fiber-reinforced silica aerogel composite. *Mater. Sci. Eng. A* **2011**, *528*, 4830–4836. [CrossRef]

25. Chakraborty, S.; Pisal, A.A.; Kothari, V.K.; Rao, A.V. Synthesis and Characterization of Fibre Reinforced Silica, Aerogel Blankets for Thermal. *Adv. Mater. Sci. Eng.* **2016**, *2016*. [CrossRef]

26. Li, Z.; Gong, L.; Cheng, X.; He, S.; Li, C.; Zhang, H. Flexible silica aerogel composites strengthened with aramid fibers and their thermal behavior. *Mater. Des.* **2016**, *99*, 349–355. [CrossRef]

27. Shafi, S.; Navik, R.; Ding, X.; Zhao, Y. Improved heat insulation and mechanical properties of silica aerogel/glass fiber composite by impregnating silica gel. *J. Non-Cryst. Solid* **2019**, *503–504*. [CrossRef]

28. Zhu, L.; Wang, Y.; Cui, S.; Yang, F.; Nie, Z.; Li, Q.; Wei, Q. Preparation of silica aerogels by ambient pressure drying without causing equipment corrosion. *Molecules* **2018**, *23*, 1935. [CrossRef]

29. Zhou, T.; Gong, L.; Cheng, X.; Pan, Y.; Li, C.; Zhang, H. Preparation and characterization of silica aerogels from by-product silicon tetrachloride under ambient pressure drying. *J. Non-Crystt. Solid* **2018**, *499*, 387–393. [CrossRef]

30. Hasan, M.A.; Rashmi, S.; Esther, A.C.M.; Bhavanisankar, P.Y.; Sherikar, B.; Sridhara, N.; Dey, A. Evaluations of silica aerogel-based flexible blanket as passive thermal control element for spacecraft applications. *J. Mater. Eng. Perform.* **2018**, *27*, 1265–1273. [CrossRef]

31. Lakatos, Á. Thermal conductivity of insulations approached from a new aspect. *J. Therm. Anal. Calorim.* **2018**, *133*, 329–335. [CrossRef]

32. EN ISO 12664. *Thermal Performance of Building Materials and Products. Determination of Thermal Resistance by Means of Guarded Hot Plate and Heat Flow Meter Methods*; Dry and moist products of medium and low thermal resistance, CEN–EN–Standard; BSI: London, UK, 2001.

33. ASTM C518-17. *Standard Test Method for Steady-State Thermal Transmission Properties by Means of the Heat Flow Meter Apparatus*; ASTM Internationa: West Conshohocken, PA, USA, 2017.

34. ISO 8301. *ISO 8301:1991 Thermal Insulation Determination of Steady-State Thermal Resistance and Related Properties Heat Flow Meter Apparatus*; American National Standards Institute (ANSI): Washington, DC, USA, 1991.

35. Lakatos, Á. Method for the determination of sorption isotherms of materials demonstrated through soil samples. *Int. Rev. Appl. Sci. Eng.* **2011**, *2*, 117–121. [CrossRef]

36. Lakatos, Á.; Deák, I.; Berardi, U. Thermal characterization of different graphite polystyrene. *Int. Rev. Appl. Sci. Eng.* **2018**, *9*, 163–168. [CrossRef]

37. Lakatos, Á.; Trnik, A. Thermal characterization of fibrous aerogel blanket. *AIP Conf. Proc.* (accepted).

38. EN ISO 12571. *Hygrothermal Performance of Building Materials and Products—Determination of Hygroscopic Sorption Properties*; BSI: Dublin, Ireland, 2013.

39. DIN 51007. *General Principles of Differential Thermal Analysis*; Deutsches Institut für Normung, DIN: Berlin, Germany, 1994.

40. EN ISO 12667. *Thermal Performance of Building Materials and Products. Determination of Thermal Resistance by Means of Guarded Hot Plate and Heat Flow Meter Methods*; Products of high and medium thermal resistance; BSI: London, UK, 2001.

41. Brunauer, S.; Deming, L.S.; Deming, W.E.; Teller, E. On a theory of the van der Waals adsorption of gases. *J. Am. Chem. Soc.* **1940**, *62*, 1723–1732. [CrossRef]

42. Liang, Y.; Wu, H.; Huang, G.; Yang, J.; Wang, H. Thermal performance and service life of vacuum insulation panels with aerogel composite cores. *Energy Build.* **2017**, *154*, 606–617. [CrossRef]

43. Musić, S.; Filipović-Vinceković, N.; Sekovanić, L. Precipitation of amorphous SiO_2 particles and their properties. *Braz. J. Chem. Eng.* **2011**, *28*, 89–94. [CrossRef]

44. Kwon, Y.G.; Choi, S.E.Y. Ambient-dried silica aerogel doped with TiO_2 powder for thermal insulation. *J. Mater. Sci.* **2000**, *35*, 6075–6079. [CrossRef]

Energies **2019**, *12*, 2001

45. Zhang, H.; Zhang, C.; Ji, W.; Wang, X.; Li, Y.; Tao, W. Experimental characterization of the thermal conductivity and microstructure of opacifier-fiber-aerogel composite. *Molecules* **2018**, *23*, 2198. [CrossRef] [PubMed]
46. Kunic, R. Vacuum insulation panels—an assessment of the impact of accelerated ageing on service life. *Strojniški Vestnik—J. Mech. Eng.* **2012**, *58*, 598–606. [CrossRef]

energies

MDPI

Article

Preparation and Characterization of Novel Plaster with Improved Thermal Energy Storage Performance

Jan Fořt [1,2,*], Radimír Novotný [1], Anton Trník [2,3] and Robert Černý [2]

[1]   Institute of Technology and Business in České Budějovice, Okružní 517/10, 370 01 České Budějovice, Czech Republic
[2]   Department of Materials Engineering and Chemistry, Faculty of Civil Engineering, Czech Technical University in Prague, Thákurova 7, 16629 Prague 6, Czech Republic
[3]   Department of Physics, Faculty of Natural Sciences, Constantine the Philosopher University in Nitra, A. Hlinku 1, 94974 Nitra, Slovakia
*   Correspondence: jan.fort@fsv.cvut.cz

Received: 25 July 2019; Accepted: 26 August 2019; Published: 28 August 2019

Abstract: Thermal energy storage systems based on latent heat utilization represent a promising way to achieve building sustainability and energy efficiency. The application of phase change materials (PCMs) can substantially improve the thermal performance of building envelopes, decrease the energy consumption, and support the thermal comfort maintenance, especially during peak periods. On this account, the newly formed form-stable PCM (FSPCM) based on diatomite impregnated by dodecanol is used as an admixture for design of interior plasters with enhanced thermal storage capability. In this study, the effect of FSPCM admixture on functional properties of plasters enriched by 8, 16 and 24 wt.% is determined. On this account, the assessment of physical, thermal, hygric, and mechanical properties is done in order to correlate obtained results with applied FSPCM dosages. Achieved results reveal only a minor influence of applied FSPCM admixture on material properties when compared to negative impacts of commercially produced PCMs. The differential scanning calorimetry discloses variations of the phase change temperature, which ranging from 20.75 °C to 21.68 °C and the effective heat capacity increased up to 15.38 J/g accordingly to the applied FSPCM dosages.

Keywords: phase change temperature; plaster; thermal energy storage; mechanical properties; thermal properties

---

1. Introduction

Nowadays, the rising energy consumption related to the building sector is a source of concern associated with the energy inefficiency and excessive production of greenhouse gases (GHG). Namely, the buildings sector consumes about 40% of annually produced primary energy in the European Union and is responsible for the production of almost 25% of greenhouse gases (GHG) [1]. Another important issue is linked with the rise of fuel prices and abandonment of conventional energy sources, including the combustion of fossil fuels which are viewed as one of the major sources of excessive emissions [2]. However, despite several developed mitigation strategies aimed at improvements of the energy performance, the achieved results and observations cannot be considered satisfactory in terms of sustainable development goals.

Presently, the utilization of traditional insulation materials such as polystyrene, mineral wool or polyurethane foam have reached the maximal insulation potential. Thus the thermal stability of buildings can be upgraded only by advanced energy improvement provisions or combination of several strategies [3]. The recent development of innovative insulation materials such as vacuum insulation panels, gas-filled panels or aerogels represents a very efficient method for the energy performance of building envelopes, however despite the superior thermal resistance of mentioned materials, ambitious

GHG mitigation targets require the employment of additional methods [3–5]. As one of the most critical parameters can be seen in a low thermal inertia of current insulation materials and consequent limited thermal stability of building envelopes during peak periods [6]. Moreover, considering the recent knowledge related to the energy efficiency of buildings and oncoming climate change [7], requirements on interior air quality maintenance are going to be more important in the near future, when the peak temperatures will be probably increased.

Advanced systems of the thermal energy storage have been extensively studied to meet energy efficiency criteria with increased demands on the maintenance of indoor thermal comfort of modern residential and office buildings [8]. Thermal energy storage systems have been found to be a beneficial step towards balancing the energy availability and energy demand period for heating and cooling [9]. Specifically, the passive cooling/heating technology based on the utilization of latent heat could substantially contribute to the preservation of ambient thermal comfort and consequently reduce the costs associated with energy consumption. The high potential of PCMs lays especially in the effectivity during daily temperature variations and subsequent mitigation of indoor temperature swings. Considering the application of PCMs in the construction sector, several approaches contemplating PCM incorporation into building elements can be identified.

However, several limitations for passive heating/cooling PCM-based systems have been revealed [10]. Since the passive systems completely depend on fluctuations of outdoor temperature, identification of ideal PCM having a suitable temperature operating range is required to drop/overcome the solidification/melting point. Otherwise, inappropriately selected PCMs cannot serve as cooling/heating support and the efficiency of the system is limited. This problem can be solved by the utilization of addition HVAC devices, nevertheless, their use lessens the gained benefits. As is evident, a proper understanding and determination of PCMs properties poses an essential knowledge base for thermal energy system application and promotion of their great potential [11].

The temperature moderation capability of PCMs has attracted the attention of several researchers intending to develop concrete with improved thermal storage properties [12,13]. However, according to the restriction related to the negative effect of incorporated PCMs on material properties, a major barrier can be viewed in PCMs compatibility with cementitious materials, which limits broader PCM applications [14]. A subsequent optimization of the applied dosages of PCMs in order to increase the thermal storage capacity improved the efficiency up to 35%, which resulted in decreased overheating rate as well as in increased economic viability [15].

Another substantial obstacle for a broader application of thermal energy storage is the negative effect of commercially produced polymer-based PCMs on the functional properties of modified materials. Despite the beneficial effect of PCMs on preservation of ambient climate with reduced energy consumption, the PCMs incorporation into cementitious materials is accompanied by many issues such as abrasion of PCM shells, clumping or segregation of PCM particles, and leakage of PCM media [12,16]. Especially, notable leakage together with a poor compatibility with cementitious materials represent major barriers for PCM utilization. Minor improvements in the field of microencapsulated PCM applications were presented by Zhang et al. [17] and Sari et al. [18], however attempts aimed at fabrication of form-stable PCMs can be viewed as major breakthrough. The impregnation method based on the absorption of paraffin/salt by a suitable bearer with a highly porous structure has been reported as a promising solution overcoming the problems described above. For example, He et al. [19] have impregnated perlite with a combination of eutectic PCMs for application in cement mortars. These efforts resulted in the improvement of the thermal storage properties while the mechanical strength of fabricated PCMs was reduced only slightly and could meet the technical requirements for building mortars. It was found that the thermal and mechanical properties of mortars with PCMs are highly dependent on the porosity of the mortars [20]. In other words, the physical and chemical compatibility between particular materials represents a crucial factor for the effective design of PCM mortars [21].

Taking into account the abovementioned points, the development and experimental analysis of newly developed form-stable PCM (FSPCM) incorporated into plaster mixture were carried out. In this paper the following aspects were considered: the suitability of the developed FSPCM for incorporation into cement-lime plaster; the influence of the applied FSPCM admixture on the material properties of prepared plasters, and sufficient thermal energy storage capacity. Material functional performance is determined by the mean basic physical, mechanical, thermal, and hygric properties. Particular importance is given to the improvement and long-term stability of the thermal storage properties. The developed plasters can be easily applied as an additional interior layer to improve the thermal performance of light-weighted building envelopes with poor thermal inertia.

2. Materials and Methods

2.1. Studied Materials

The FSPCM was fabricated on the basis of a vacuum saturation principle using highly porous diatomite powder (LB Minerals, Horní Bříza, Czech Republic) together with *n*-dodecanol (Sigma-Aldrich, Taufkirchen, Germany) as a PCM medium. According to material characteristics provided by the producer; the phase change temperature is about 22 °C and the latent heat about 170 J/g. Dodecanol was mixed with diatomite particles in the selected ratio (about 0.85/1 according to pozzolanic test results) and placed in a VakuCell vacuum oven (BMT Medical Technology, Brno, Czech Republic. Both materials were subsequently heated, thus melted dodecanol impregnated the porous structure of diatomite particles. The obtained mixture was milled to crush clumped particles and obtain a powdered material. The whole procedure was repeated to ensure the uniform distribution of the dodecanol in the diatomite particles. After three cycles of vacuum saturation, the prepared FSCPM did not exhibit any leakage of liquid dodecanol during heating cycles thanks to the capillary and surface tension forces [22]. The utilization of diatomite as PCM bearer preserved the pozzolanic properties up to 0.85/1 dodecanol diatomite ratio, based on the results of a Frattini test specified by the European standard ČSN EN 196-5 [23]. Such results predetermined the suitability for application in cementitious composites. The phase change temperature of the developed FSPCM was 23.15 °C during heating and 21.13 °C during cooling. The phase change enthalpy reached 71.36 J/g during cooling and 73.1 J/g during the heating cycle [24].

The particle size distribution (PSD) of pure diatomite, reference cement-lime plaster, and the newly developed FSPCM was measured on an Analysette 22 Micro Tec plus device (Fritsch, Northamptonshire, United Kingdom, Figure 1) working on a laser diffraction principle. The measuring range of the applied apparatus covers particle sizes from 0.08 μm up to 2000 μm. A green laser is used for the small particle range, whereas an infrared laser is utilized for the measurement of larger particles. The repeatability of the device according to ISO 13320 [25] is at $d_{50} \leq 1\%$.

Designed plasters were prepared with a commercially available dry plaster mixture (Manu 1, Baumit, Dětmarovice Czech Republic) composed of hydrated lime, cement, sand with a maximal grain size about 1 mm and additives. Selected dry plaster mixture was modified by FSPCM admixture using weight dosages of about 8, 16 and 24 wt.%. Due to a different surface area and particle size of FSPCM compared to dry plaster mixture, the water dosages needed to be adjusted to maintain the same workability of prepared mixtures. Here, the workability was verified using the flow table test having the spread diameter was approx. 180 mm in both perpendicular directions. The composition of the studied materials is given in Table 1.

Cast samples were stored for 28 days in a highly humid environment. After curing, all samples were dried at 60 °C for 48 h until constant mass and subjected to further experimental procedures.

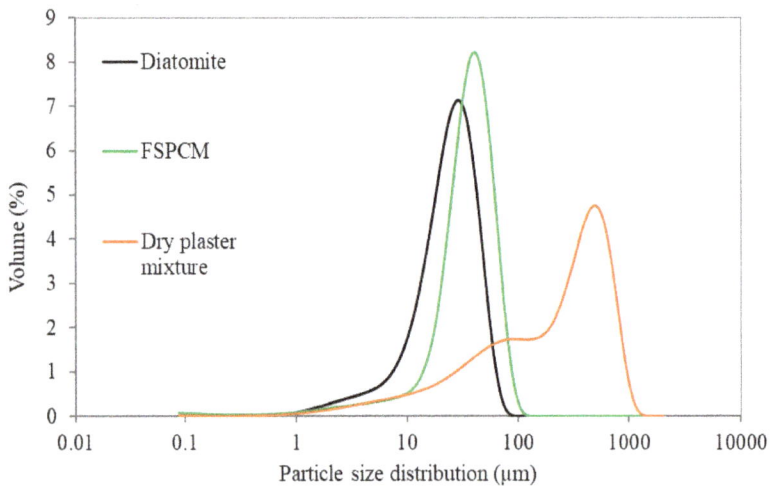

Figure 1. Particle size distribution.

Table 1. Composition of the studied plasters.

Mixture	Water	Dry Plaster	PCM	Flow Diameter
	(kg)	(kg)	(kg)	(mm)
Reference plaster (RP)	1.5	6.3	0	180
Plaster with 8 wt.% of FSPCM (P8)	1.75	6.3	0.5	177
Plaster with 16 wt.% of FSPCM (P16)	2.05	6.3	1	181
Plaster with 24 wt.% of FSPCM (P24)	2.25	6.3	1.51	183

2.2. Determination Methods

The basic physical properties of developed plasters were characterized by bulk density, matrix density, and total open porosity measurements. All these measurements were done on five cubic samples wide sides of about 50 mm. The bulk density was determined based on the gravimetrical principle (using a digital caliper and weights). The matrix density was obtained by a Pycnomatic ATC helium pycnometer (Thermo Fisher Scientific, Waltham, Massachusetts, USA).

Determination of the pozzolanic activity was done by performing a Frattini test, specified by the European standard ČSN EN 196-5 [23]. The tested specimens were mixed with 100 mL of boiled distilled water. Afterward, samples were placed in a sealed plastic bottle and dried at 40 °C for 8 days and then were filtered with a Buchner funnel. The concentration of OH^- ions was analyzed by titration against HCl with bromophenol blue indicator, and for Ca^{2+} concentration by pH adjustment to 12.5 by NaOH, followed by titration with ethylenediaminetetraacetic acid (EDTA) solution using Murexide indicator [24].

Characterization of the inner structure of designed plasters was done by Mercury Intrusion Porosimetry (MIP) analysis. For this analysis, a combination of Pascal 140 and Pascal 440 porosimeters (Thermo Fisher Scientific, Waltham, Massachusetts, USA) was employed. The contact surface tension of mercury was 480 mN/m with the density of about 13,541 g/cm^3. The measurements were carried out at 21 °C.

Mechanical parameters such as the flexural strength and compressive strength were determined in order to access the durability of developed plasters. The measurement of compressive and flexural strengths was done by a VEB WPM Leipzig hydraulic testing device (WPM Leipzig, Leipzig, Germany) with a stiff loading frame with the capacity of 3000 kN on prismatic samples with dimensions of 40 mm × 40 mm × 160 mm.

The water vapor transmission properties of the developed plasters were measured using the cup method. Five samples with a circular cross-section of 110 mm diameter were used. The sample thickness was approx. 30 mm. The measurements were carried under isothermal conditions at a temperature of 21 °C. The sealed cup was placed in a controlled climate chamber of approx. 50% RH and weighed periodically. The steady-state values of mass gain or mass loss were utilized for the determination of the water vapor transfer properties [26].

For measurement of sorption and desorption isotherms, a dynamic DVS-Advantage vapor sorption device (Surface Measurement Systems, London, United Kingdom) was used. The measurements were done at 21 °C for relative humidity levels of 20, 40, 60, 80 and 95% RH. First, all tested samples were dried in an oven and put into a desiccator. Afterward, selected sample was placed in the climatic chamber of the DVS-Advantage device equipped by highly precise balances with resolution of 1.0 μg. A principle of the measurement is based on the gravimetric determination of mass gains and losses according to changes in relative humidity levels maintained by the device [26].

The thermal conductivity and thermal diffusivity were obtained by a hand-held portable instrument (ISOMET 2114, Applied Precision, Bratislava, Slovakia) based on a dynamic measurement principle which allow fast measurement with the accuracy of 5% of reading +0.001 W/(mK). The reproducibility of measurements is 3% when reading in the temperature range from 0 to 40 °C. Five cubic samples of about 70 mm side length were used and the obtained results were averaged.

For the measurement of phase change temperatures and enthalpies, differential scanning calorimetry (DSC) analysis was done. For this purpose, a DSC 822e apparatus (Mettler Toledo, Greifensee, Switzerland) equipped with a FT 900 cooling device (Julabo, Seelbach, Germany) was employed. During the measurements, the following temperature regimes were used: 5 min of isothermal regime, cooling at 0.5 °C/min from a temperature of 40 °C to a temperature of 0 °C, 5 min of isothermal regime, heating at 0.5 °C/min from 0 °C to 40 °C, 5 min of isothermal regime. To obtain and present reliable results, DSC measurements were performed multiple-times (5×) and consequently averaged. The long-term stability testing of the studied plasters was conducted by accelerated aging test and passing 100 cycles of the same heating/cooling rate and temperature range as described above.

3. Results

3.1. Basic Physical Properties

A material characterization of the plasters enriched by FSPCM according to their basic material properties is given in Table 2.

Table 2. Basic physical properties.

Material	Bulk Density (kg/m^3)	Matrix Density (kg/m^3)	Total Open Porosity (%)
RP	1572	2415	34.9
P8	1448	2294	37.8
P16	1386	2179	35.3
P24	1321	1982	35.6

The obtained results revealed only minor changes in the material microstructure and a slight decrease in the bulk and matrix density was noted. As one can see, the average bulk density dropped from an initial value of 1572 kg/m^3 (RP) to 1321 kg/m^3 (P24) along with the matrix density. On the other hand, the total open porosity did not reveal a sharp dependency between the amount of used FSPCM and the pore volume. While P8 plaster exhibited an increase in the pore volume of about 3%, applied higher dosages of FSPCM (16% and 24%) reduced the total open porosity compared to P8 at almost the same level as was obtained for reference plaster. The explanation of this confusing fact can be found in the pore size distribution curves plotted in Figure 2. Here, a shift in macropore range revealed for P8 plaster can be associated with changes in the workability of fresh mixtures as well as different rheological properties. Concurrently, very fine FSPCM particles apparently filled large pores and promoted the increase of pores in the range from 0.01 μm to 1 μm. As visible in Figure 2, plasters with higher dosages of FSPCM (P16 and P24) exhibit a lower threshold pore diameter compared to RP and P8 plaster. The achieved results pointed to a better incorporation of FSPCM compared to commercially produced inert PCMs [27], and according to the results revealed by Lee et al. [28] preservation of mechanical parameters can be expected.

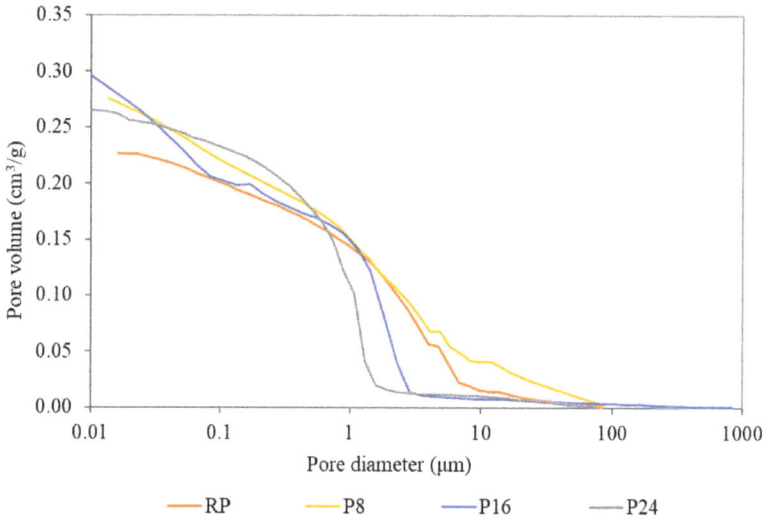

Figure 2. Pore diameter distribution of the studied plasters.

3.2. Mechanical Properties

One of the primary purposes related to the development of FSPCMs consists in the mitigation of the negative effects of commercially produced microencapsulated PCMs on the mechanical strength of designed composite materials, as described in several works [12,13,20]. The abrasion and possible damage of the polymer shells used during the mixing process represent important factors limiting a broader utilization of encapsulated PCMs in material design practice. Experimentally accessed results given in Table 3 show an effect of particular FSPCM dosages on the mechanical parameters of the tested plasters.

Table 3. Mechanical properties of studied plasters with FSPCM.

Mixture	Compressive Strength (MPa)	Flexural Strength (MPa)	Young's Modulus (GPa)
PR	1.91	0.9	3.02
P8	1.8	0.83	2.72
P16	1.73	0.78	2.69
P24	1.69	0.73	2.52

All determined mechanical parameters (Young´s modulus, compressive and flexural strength) for P24 dropped proportionally by approx. about 15% in comparison to RP. While the compressive strength was reduced from 1.91 MPa to 1.69 MPa, the flexural strength decreased only by 0.17 MPa to a final value of 0.73 MPa. Moreover, the P8 and P16 mixtures were affected in a lesser extent, thus the obtained results point to successful incorporation of FSPCM into cement-lime plaster matrix. The utilization of the newly developed FSPCM based on dodecanol and diatomite proved its contribution for the preservation of mechanical properties. Results obtained by Sun and Wang [29] for PCM composite based on paraffin and expanded perlite exhibited lower values, even when compared to P24 plaster with the highest FSCM dosage. They found that 30% of incorporated paraffin/expanded perlite composite exhibited a decrease in compressive strength of more than about 25%. This fact was caused by a high content of applied paraffin and limited reactivity of the shape stabilized composite with the cement during the hydration period.

3.3. Thermal Properties and Energy Storage

The thermal conductivity of plasters represents an important parameter for building designers, especially in the case of PCM composites. The results provided by Karkri et al. [30] highlighted that a lowered coefficient of thermal conductivity of the support material matrix is the limiting factor for PCM utilization in building practice.

Obtained thermal diffusivity and thermal conductivity results are presented in Table 4 to point out the changes induced by FSPCM admixture. As can be clearly seen, the coefficient of thermal conductivity was lowered very slightly, while only minor changes were also observed for the thermal diffusivity. This finding can be attributed to limited changes in material porosity and increased thermal conductivity of a pure FSPCM compared to commercial PCMs [31]. To be specific, almost insignificant changes in the tested thermal properties compared to the reference plaster were obtained, and the thermal conductivity values ranged between 0.52 W/(mK) and 0.54 W/(mK). The results achieved with our modified plasters are more favorable compared to the achievements discussed in the study of Kusama and Ishidoya [32], where a significant decrease in thermal conductivity was revealed for mixtures having a higher content of Micronal PCM.

Table 4. Thermal conductivity of studied plasters.

Material	λ (W/mK)	a (m^2/s)
RP	0.54	0.36
P8	0.52	0.33
P16	0.53	0.34
P24	0.53	0.34

Contrary to the conventional building practice, the reduced thermal conductivity of PCM composites cannot be viewed as beneficial according to the conclusions of Joseph et al. [33] who stated a low thermal conductivity was a barrier for sufficient thermal energy transfer from the material to the environment. As reported by Ascione [34], the insufficient or very low thermal conductivity can be a reason for limited effectivity of applied PCMs or even restrict the occurrence of the phase change [32–34]. In the light of these findings, several studies suggested the application of expanded graphite in order to increase the thermal conductivity of PCMs [17,18].

To investigate the influence of incorporated dodecanol/diatomite composite in cement-lime plaster, DSC analysis was employed for the identification of phase change intervals. The temperature dependent effective heat capacity curves obtained using a DSC device controlled by STAR SW 9.1 during the measurement of the studied modified plasters are shown in Figures 3 and 4. Here, the phase change temperatures of the studied plasters were detected in the range of 21–19 °C during cooling, while the temperature ranged from 21 °C to 26 °C during the heating cycle. While the curve of temperature-dependent effective heat capacity during heating is more widespread, the crystallization during the cooling process exhibited a higher rate, therefore the onset and endset temperature differ only by about 2 °C at maximum. Detailed information about the onset and endset temperatures, together with the heats of phase change is given in Table 5.

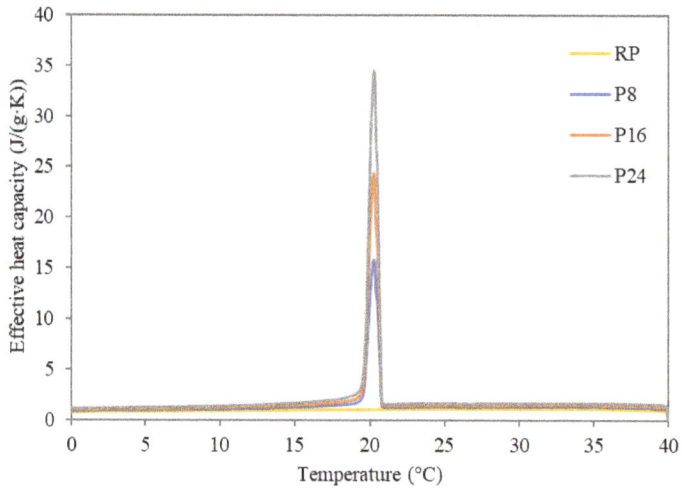

Figure 3. Temperature dependent effective heat capacity of studied FSPCM plasters during cooling.

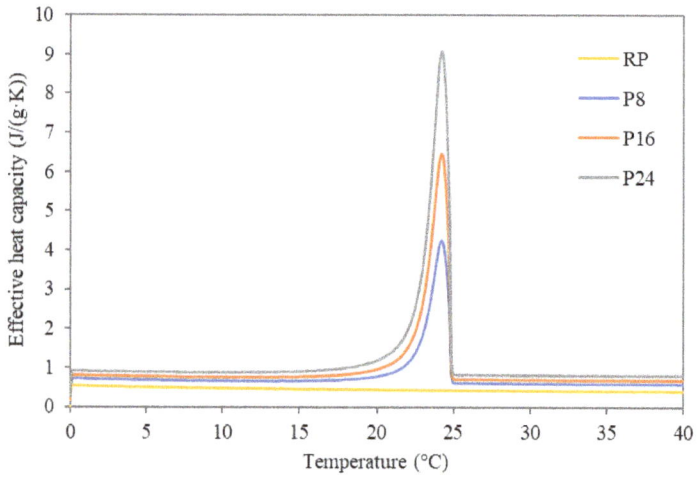

Figure 4. Temperature dependent effective heat capacity of studied FSPCM plasters during heating.

Table 5. Phase change temperatures of studied plasters and specific enthalpies.

Material	Phase Change Temperature (°C)				Phase Change Enthalpy (J/g)	
	Heating		Cooling		Heating	Cooling
	Onset	Endset	Onset	Endset		
RP	-	-	-	-	-	-
P8	21.46	24.68	20.68	18.63	4.63	4.89
P16	21.13	24.91	20.81	18.98	10.11	10.28
P24	21.68	25.86	20.75	19.66	15.20	15.38

Looking at the data, the incorporation of FSPCM into the plaster matrix did not dramatically affect the onset nor endset temperatures compared to data for pure FSPCM. Only a slight spread of the phase change interval was observed, which cannot be perceived as a barrier for material utilization.

Specifically, all measured phase change temperature intervals remained in the desired range suitable for the moderation of the indoor climate. The latent heat of prepared plasters was increased proportionally to applied FSPCM dosages, thus providing thermal storage capacity improvements compared to the reference plaster. The obtained heat flux curves kept their unimodal shape, which provides substantial benefits compared to composites based on commercial PCMs. Reported bimodal heat flux curves pointed out a reduced ability of applied PCMs to maintain the indoor temperature in the desired range due to incomplete phase changes [35]. Considering the relatively narrow operational range (desired from 20 °C to 24 °C) of the PCMs used, unimodal shape of the temperature-dependent apparent heat capacity allows maximization of the heat absorption/release effectivity. In other words, such a material ensures operation in the desired range with improved effectivity compared to bimodal shape curves with usually more widespread operational ranges. The achieved results inevitably exhibited a strong dependence between the values of latent heat and applied FSPCM dosages. P8 plaster revealed the ability to absorb about 4.63 J/g during the heating cycle and release of about 4.89 J/g during the cooling cycle. On the contrary, the plaster with the highest FSPCM dosage showed better results, namely a phase change enthalpy of about 15.20 J/g during heating and 15.38 J/g during the cooling cycle. Considering the results revealed in similar studies, the developed plaster with FSPCM exhibited a notable improvement compared to similar products. Namely, in the work of Xu and Li [36] about three times lower values (ranging from 1.32 J/g to 5.44 J/g) in cement composite enriched by paraffin-based form-stable PCM were recorded. The attempt of Liu et al. [37] to develop calcium silicate-coated expanded clay-based form stable PCM for application in cement composite delivered a substantial improvement (about 58% increase in effective heat capacity compared to reference sample). Nonetheless, the latent heat varied from 1.54 J/g to 2.8 J/g only. From this point of view, the developed composite plasters provide substantially improved thermal storage properties in the desired temperature range.

3.4. Long-Term Stability of FSPCM Plaster Thermal Storage Properties

The long-term stability of FSPCM poses another critical parameter for assessment of thermal energy storage composites. In order to determine this parameter, P24 plaster with the highest content of FSPCM (highest probability of leakage) was studied by DSC analysis intermediately after fabrication and after completing 100 cycles (see Figure 5).

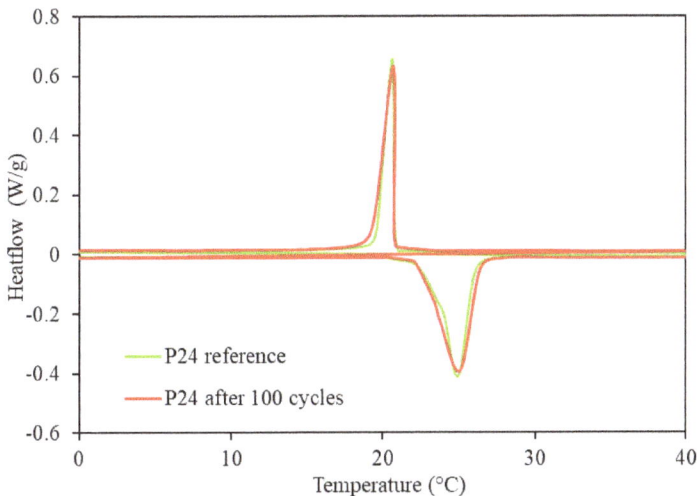

Figure 5. Long-term stability of developed plaster—mixture P24 after 100 cycles.

Looking at the plotted curves, the onset and endset temperatures did not display any distinct shift or drop. The onset phase change temperature during cooling was moved slightly from 20.69 °C to 20.75 °C and from 21.53 °C to 21.68 °C during heating. The latent heat of fusion decreased from 15.78 to 15.20 J/g, and the latent heat of freezing was reduced from 15.89 J/g to 15.38 J/g. Therefore, this minor drop clearly depicted the long-term stability and no leakage was observed. As described by Feczko et al. [35], the leakage is usually recognized within the first cycles and decrease in time (higher number of cycles). Taking into account these results, the FSPCM plasters based on dodecanol and diatomite remain stable even after 100 thermal cycles and can be expected to preserve this stability even after a higher number of cycles.

3.5. Hygric Properties

The water vapor resistance factors determined by the cup method, together with the moisture conductivity of studied FSPCM plasters are given in Table 6.

Table 6. Hygric properties of the studied plasters.

Material	κ (m^2/s)	μ (-)
RP	5.66×10^{-7}	9.4
P8	5.87×10^{-7}	8.9
P16	6.23×10^{-7}	8.5
P24	6.80×10^{-7}	8.3

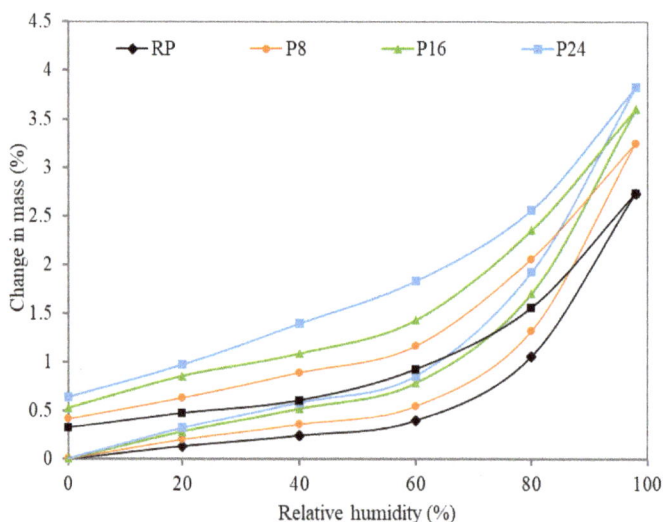

Figure 6. Sorption and desorption isotherms.

These parameters describe the water transport predominantly from the exterior side to the interior and vice versa, which substantially affects the thermal conductivity. As is clearly visible, the obtained experimental results did not indicate any substantial differences between reference samples and modified plasters. While the water vapor resistance factor for RP reached 9.4, the applied FSPCM moved this value for P8, P16, and P24 to 8.9, 8.5 and 8.3, respectively. The calculated parameters exhibited only a slight decrease in the water vapor resistivity which can be partially assigned to changes in the porous space, the formation of pores with smaller diameter compared to reference plaster and on the other hand to the improved hygroscopicity. This effect is evident from Figure 6, where the sorption and desorption isotherms are plotted. Here, studied samples were exposed to various levels of relative humidity ranging from 0 to approx. 95% at a constant temperature (21 °C) to reveal the relationship between sample mass and the applied relative humidity level.

The equilibrium moisture content (EMC) can also be used for interpretation of material ability to moderate indoor humidity during the diurnal swings. The record of this measurement complies with previously described results of water vapor transmission properties. The EMC of RP, P8, P16, and P24 was measured to be 2.74, 3.24, 3.60 and 3.83%, respectively. Obtained results of modified plasters represent a distinct shift in the measured parameter, which can be beneficially utilized for control of indoor humidity levels in terms of moisture buffering [38].

4. Conclusions

This study focused on the design of new thermal energy storage plasters provided detailed information about the basic physical, thermal, mechanical, and hygric properties of modified plasters enhanced by 8, 16, and 24 wt.% of FSPCM based on diatomite and dodecanol. The obtained results represent important data for the applicability of the studied materials as well as moderation of the indoor climate with respect to the diurnal or seasonal temperature fluctuations or changes. The following points should be highlighted:

- A substantial improvement was achieved in the field of preservation of functional parameters, where only slight deterioration was found. Namely, a decrease of about only 13% for the compressive and flexural strengths was observed compared to reference plaster in the case of a 24 wt.% FSPCM admixture. This finding showed that diatomite particles could be utilized as suitable bearer material for PCMs due to their compatibility with cementitious materials. Moreover, the developed FSPCM exhibited improved mechanical performance compared to commercially produced PCMs.
- Thermal conductivity of the developed plasters was decreased only slightly, therefore the capability to release and absorb thermal energy was influenced only slightly.
- Hygric parameters of the modified plaster were slightly improved thanks to an increase in the total open porosity, thus the material provides a better moisture-moderation capability compared to reference plaster.
- The incorporation of FSPCM into the cement-lime plaster did not shift the phase change temperature, which remained stable for all tested mixtures. On the other hand, the phase change enthalpy was proportionally decreased depending on the volume of the applied PCM. The phase change temperatures varied from 21.68 °C during heating to 20.75 °C during cooling for plaster with 24 wt.% of FSPCM, while the phase change enthalpy reached values of about 15.38 J/g during cooling, and 15.20 J/g during heating, respectively. The obtained values remained stable even after 100 heating-cooling cycles, therefore, no leakage was detected.

Considering the results of this study, further experimental tests for verification of the positive effect of the form-stable PCM composite should be performed. Here, semi-scale and full-scale experiments in particular could be helpful in order to prove the assumed energy savings in building maintenance.

Author Contributions: Conceptualization, J.F. and R.Č.; methodology, J.F: and A.T.; validation, R.N. and A.T.; formal analysis, R.Č.; investigation, J.F., R.N. and A.T.; resources, R.N and R.Č.; data curation, J.F and A.T.; writing—original draft preparation, J.F.; writing—review and editing, J.F.; visualization, J.F.; supervision, R.Č.; project administration, J.F.; funding acquisition, J.F. and R.N.

Funding: This research has been supported by the Czech Science Foundation, under project No 18-03997S and specific university research of the Institute of Technology and Business in České Budějovice, under project No SVV201906.

Conflicts of Interest: The authors declare no conflict of interest.

References

1. Samuel, D.G.L.; Nagendra, S.M.S.; Maiya, M.P. Passive alternatives to mechanical air conditioning of building: A review. *Build. Environ.* **2013**, *66*, 54–64. [CrossRef]
2. Allwood, J.M.; Cullen, J.M.; Milford, R.L. Options for Achieving a 50% Cut in Industrial Carbon Emissions by 2050. *Environ. Sci. Technol.* **2010**, *44*, 1888–1894. [CrossRef] [PubMed]
3. Jelle, B.P. Traditional, state-of-the-art and future thermal building insulation materials and solutions—Properties, requirements and possibilities. *Energy Build.* **2011**, *43*, 2549–2563. [CrossRef]
4. Alam, M.; Singh, H.; Suresh, S.; Redpath, D.A.G. Energy and economic analysis of Vacuum Insulation Panels (VIPs) used in non-domestic buildings. *Appl. Energy* **2017**, *188*, 1–8. [CrossRef]
5. Hill, C.; Norton, A.; Dibdiakova, J. A comparison of the environmental impacts of different categories of insulation materials. *Energy Build.* **2018**, *162*, 12–20. [CrossRef]
6. Pombo, O.; Allacker, K.; Rivela, B.; Neila, J. Sustainability assessment of energy saving measures: A multi-criteria approach for residential buildings retrofitting-A case study of the Spanish housing stock. *Energy Build.* **2016**, *116*, 384–394. [CrossRef]
7. Koci, J.; Koci, V.; Madera, J.; Cerny, R. Effect of applied weather data sets in simulation of building energy demands: Comparison of design years with recent weather data. *Renew. Sustain. Energy Rev.* **2019**, *100*, 22–32. [CrossRef]
8. Alva, G.; Lin, Y.X.; Liu, L.K.; Fang, G.Y. Synthesis, characterization and applications of microencapsulated phase change materials in thermal energy storage: A review. *Energy Build.* **2017**, *144*, 276–294. [CrossRef]
9. Bruno, F.; Tay, N.H.S.; Belusko, M. Minimising energy usage for domestic cooling with off-peak PCM storage. *Energy Build.* **2014**, *76*, 347–353. [CrossRef]
10. Park, J.H.; Jeon, J.; Lee, J.; Wi, S.; Yun, B.Y.; Kim, S. Comparative analysis of the PCM application according to the building type as retrofit system. *Build. Environ.* **2019**, *151*, 291–302. [CrossRef]
11. Ramakrishnan, S.; Wang, X.M.; Sanjayan, J.; Petinakis, E.; Wilson, J. Development of thermal energy storage cementitious composites (TESC) containing a novel paraffin/hydrophobic expanded perlite composite phase change material. *Sol. Energy* **2017**, *158*, 626–635. [CrossRef]
12. Biswas, K.; Lu, J.; Soroushian, P.; Shrestha, S. Combined experimental and numerical evaluation of a prototype nano-PCM enhanced wallboard. *Appl. Energy* **2014**, *131*, 517–529. [CrossRef]
13. Figueiredo, A.; Lapa, J.; Vicente, R.; Cardoso, C. Mechanical and thermal characterization of concrete with incorporation of microencapsulated PCM for applications in thermally activated slabs. *Constr. Build. Mater.* **2016**, *112*, 639–647. [CrossRef]
14. Entrop, A.G.; Brouwers, H.J.H.; Reinders, A. Experimental research on the use of micro-encapsulated Phase Change Materials to store solar energy in concrete floors and to save energy in Dutch houses. *Sol. Energy* **2011**, *85*, 1007–1020. [CrossRef]
15. Hunger, M.; Entrop, A.G.; Mandilaras, I.; Brouwers, H.J.H.; Founti, M. The behavior of self-compacting concrete containing micro-encapsulated Phase Change Materials. *Cem. Concr. Compos.* **2009**, *31*, 731–743. [CrossRef]
16. Costanzo, V.; Evola, G.; Marletta, L.; Nocera, F. The effectiveness of phase change materials in relation to summer thermal comfort in air-conditioned office buildings. *Build. Simul.* **2018**, *11*, 1145–1161. [CrossRef]
17. Zhang, Z.G.; Shi, G.Q.; Wang, S.P.; Fang, X.M.; Liu, X.H. Thermal energy storage cement mortar containing n-octadecane/expanded graphite composite phase change material. *Renew. Energy* **2013**, *50*, 670–675. [CrossRef]
18. Sari, A.; Karaipekli, A. Preparation, thermal properties and thermal reliability of palmitic acid/expanded graphite composite as form-stable PCM for thermal energy storage. *Sol. Energy Mater. Sol. Cells* **2009**, *93*, 571–576. [CrossRef]
19. He, Y.; Zhang, X.; Zhang, Y.J. Preparation technology of phase change perlite and performance research of phase change and temperature control mortar. *Energy Build.* **2014**, *85*, 506–514. [CrossRef]
20. Schossig, P.; Henning, H.M.; Gschwander, S.; Haussmann, T. Micro-encapsulated phase-change materials integrated into construction materials. *Sol. Energy Mater. Sol. Cells* **2005**, *89*, 297–306. [CrossRef]
21. Figueiredo, A.; Vicente, R.; Lapa, J.; Cardoso, C.; Rodrigues, F.; Kampf, J. Indoor thermal comfort assessment using different constructive solutions incorporating PCM. *Appl. Energy* **2017**, *208*, 1208–1221. [CrossRef]
22. Fort, J.; Trnik, A.; Pavlikova, M.; Pavlik, Z.; Cerny, R. Fabrication of Dodecanol/Diatomite Shape-Stabilized PCM and Its Utilization in Interior Plaster. *Int. J. Thermophys.* **2018**, *39*, 11. [CrossRef]

23. Czech Standardization Institute. *Methods of Cement Testing—Part 5: Pozzolanicity Testing*, ČSN EN 196-5; Czech Standardization Institute: Prague, Czech Republic, 2011.

24. Fort, J.; Pavlikova, M.; Zaleska, M.; Pavlik, Z.; Trnik, A.; Jankovsky, O. Preparation of puzzolana active two component composite for latent heat storage. *Ceram. Silik.* **2016**, *60*, 291–298. [CrossRef]

25. British Standards Institution. *Particle Characterization*, ISO 13320. ISO/TC24/SC4/WG6. British Standards Institution: London, UK, 1981.

26. Fort, J.; Pavlik, Z.; Zumar, J.; Pavlikova, M.; Cerny, R. Effect of temperature on water vapor transport properties. *J. Build. Phys.* **2014**, *38*, 156–169. [CrossRef]

27. Souayfane, F.; Fardoun, F.; Biwole, P.H. Phase change materials (PCM) for cooling applications in buildings: A review. *Energy Build.* **2016**, *129*, 396–431. [CrossRef]

28. Lee, K.O.; Medina, M.A.; Sun, X.Q.; Jin, X. Thermal performance of phase change materials (PCM)-enhanced cellulose insulation in passive solar residential building walls. *Sol. Energy* **2018**, *163*, 113–121. [CrossRef]

29. Sun, D.; Wang, L.J. Utilization of paraffin/expanded perlite materials to improve mechanical and thermal properties of cement mortar. *Constr. Build. Mater.* **2015**, *101*, 791–796. [CrossRef]

30. Karkri, M.; Lachheb, M.; Nogellova, Z.; Boh, B.; Sumiga, B.; AlMaadeed, M.A.; Fethi, A.; Krupa, I. Thermal properties of phase-change materials based on high-density polyethylene filled with micro-encapsulated paraffin wax for thermal energy storage. *Energy Build.* **2015**, *88*, 144–152. [CrossRef]

31. Pavlik, Z.; Fort, J.; Pavlikova, M.; Pokorny, J.; Trnik, A.; Cerny, R. Modified lime-cement plasters with enhanced thermal and hygric storage capacity for moderation of interior climate. *Energy Build.* **2016**, *126*, 113–127. [CrossRef]

32. Kusama, Y.; Ishidoya, Y. Thermal effects of a novel phase change material (PCM) plaster under different insulation and heating scenarios. *Energy Build.* **2017**, *141*, 226–237. [CrossRef]

33. Joseph, A.; Kabbara, M.; Groulx, D.; Allred, P.; White, M.A. Characterization and real-time testing of phase-change materials for solar thermal energy storage. *Int. J. Energy Res.* **2016**, *40*, 61–70. [CrossRef]

34. Ascione, F. Energy conservation and renewable technologies for buildings to face the impact of the climate change and minimize the use of cooling. *Sol. Energy* **2017**, *154*, 34–100. [CrossRef]

35. Feczko, T.; Trif, L.; Horak, D. Latent heat storage by silica-coated polymer beads containing organic phase change materials. *Sol. Energy* **2016**, *132*, 405–414. [CrossRef]

36. Xu, B.W.; Li, Z.J. Paraffin/diatomite composite phase change material incorporated cement-based composite for thermal energy storage. *Appl. Energy* **2013**, *105*, 229–237. [CrossRef]

37. Liu, Y.S.; Xie, M.J.; Xu, E.T.; Gao, X.; Yang, Y.Z.; Deng, H.W. Development of calcium silicate-coated expanded clay based form-stable phase change materials for enhancing thermal and mechanical properties of cement-based composite. *Sol. Energy* **2018**, *174*, 24–34. [CrossRef]

38. Wu, Z.M.; Qin, M.H.; Zhang, M.J. Phase change change humidity control material and its impact on building energy consumption. *Energy Build.* **2018**, *174*, 254–261. [CrossRef]

energies

MDPI

Article

Optimising Convolutional Neural Networks to Predict the Hygrothermal Performance of Building Components

Astrid Tijskens *, Hans Janssen and Staf Roels

Department of Civil Engineering, KU Leuven, Building Physics Section, Kasteelpark Arenberg 40 Bus 2447, 3001 Heverlee, Belgium; hans.janssen@kuleuven.be (H.J.); gustaaf.roels@kuleuven.be (S.R.)
* Correspondence: astrid.tijskens@kuleuven.be; Tel.: +32-16-323-808

Received: 18 September 2019; Accepted: 16 October 2019; Published: 18 October 2019

Abstract: Performing numerous simulations of a building component, for example to assess its hygrothermal performance with consideration of multiple uncertain input parameters, can easily become computationally inhibitive. To solve this issue, the hygrothermal model can be replaced by a metamodel, a much simpler mathematical model which mimics the original model with a strongly reduced calculation time. In this paper, convolutional neural networks predicting the hygrothermal time series (e.g., temperature, relative humidity, moisture content) are used to that aim. A strategy is presented to optimise the networks' hyper-parameters, using the Grey-Wolf Optimiser algorithm. Based on this optimisation, some hyper-parameters were found to have a significant impact on the prediction performance, whereas others were less important. In this paper, this approach is applied to the hygrothermal response of a massive masonry wall, for which the prediction performance and the training time were evaluated. The outcomes show that, with well-tuned hyper-parameter settings, convolutional neural networks are able to capture the complex patterns of the hygrothermal response accurately and are thus well-suited to replace time-consuming standard hygrothermal models.

Keywords: Metamodeling; Convolutional neural networks; Time series modelling; Probabilistic assessment; Hygrothermal assessment

1. Introduction

When simulating the hygrothermal behaviour of a building component, one is confronted with many uncertainties, such as those in the exterior and interior climates, in the material properties, or even in the configuration geometry. A deterministic assessment does not enable taking into account these uncertainties, and as such, often does not allow for a reliable design decision or conclusion. A probabilistic analysis [1–6], on the other hand, enables including these uncertainties, and thus allows a more reliable assessment of the hygrothermal performance and the potential moisture damages. For this purpose, usually, the Monte Carlo approach [7] is adopted, where the uncertain input parameters' distributions are sampled multiple times and a deterministic simulation is executed for each sampled parameter combination. This approach often involves thousands of simulations and therefore, easily becomes computationally inhibitive. To surmount this problem, the hygrothermal model can be replaced by a metamodel, which is a simpler and faster mathematical model mimicking the original model, thus strongly reducing the calculation time. Static metamodels have already been applied in the field of building physics multiple times [8–10]. The main disadvantage is that these types of metamodels are developed for a specific single-valued performance indicator (e.g., the total heat loss or the final mould growth index). The wish to use a different performance indicator would require the construct of a new metamodel, which is time-intensive. Additionally, single-valued performance indicators provide less information, which might impede decision-making. For example, the maximum mould growth index is calculated based on the temperature and relative humidity time series and shows the maximum value over a period, but does not allow for assessing how long or how often this maximum occurs, or how high the mould growth index is the rest of the time.

Dynamic metamodels, on the other hand, aim to predict actual time series (temperature, relative humidity, moisture content, etc.), and thus provide a more flexible approach. Predicting the hygrothermal time series allows post-processing by any desired damage prediction model (e.g., the mould growth index), as well as provides information over the whole period. Using a metamodel to predict time series, rather than single-value performance indicators, is, to the authors' knowledge, new to the field of building physics. However, it is also more difficult, as the metamodel must be able to capture the complex and time-dependent pattern between input and output time series, and not all metamodeling strategies are suited for time series prediction.

In a previous study [11], the authors demonstrated that neural networks are well-suited to reproduce the dynamic hygrothermal response of a building component. Three popular types of neural networks were considered: multilayer perceptrons (MLP), the long-short-term memory network (LSTM) and the gated recurrent unit network (GRU), both of which are a type of recurrent neural network (RNN), and convolutional neural network (CNN). These networks were trained to predict the hygrothermal time series such as temperature, relative humidity and moisture content at certain positions in a masonry wall, based on the time series of exterior and interior climate data. The results showed that a memory mechanism to access information from past time steps is required for accurate prediction performance. Hence, only the RNN and the CNN were found to be adequate. Furthermore, the CNN was shown to outperform the RNN and was also much faster to train.

This study builds upon these previous findings. As the CNN was found to perform best, it is developed further, aiming to replace HAM-simulations (HAM: Heat, Air and Moisture) for a spectrum of facade constructions (with different geometry and materials) and/or boundary conditions (with varying exterior and interior climate, orientation, wind-driven rain, etc.). During development, many parameters inherent to the neural network architecture and training process—called the hyper-parameters—need to be defined though. Considering that these parameters can significantly influence the network's performance, it is important to choose the most optimal combination. However, this is usually a trial-and-error process, as there are no general guidelines. This paper hence proposes an approach to optimise these hyper-parameters, using the Grey-Wolf Optimisation (GWO) algorithm, as it was found competent for other applications [12,13]. This is applied to a one-dimensional (1D) brick wall, of which, the hygrothermal performance is evaluated for typical moisture damage patterns.

The next section first presents the architecture of the convolutional neural network. Next, the hyper-parameters optimisation method is explained, after which, the networks' performance evaluation is described. Section 3 describes the application and calculation object and in Section 4, the results of the hyper-parameter optimization and the networks' performance are brought together and discussed. In the conclusions, the main findings are summarised, and some final remarks are drawn.

2. Optimising Convolutional Neural Networks (CNN)

2.1. The Network Architecture

Convolutional neural networks are a class of deep neural networks most commonly applied to image analysis. More recently though, CNNs have been applied to sequence learning as well [11,14,15]. A convolution is a mathematical operation on two functions to produce a third function, defined as the integral of the product of these functions after one is reversed and shifted. In the case of a CNN, the convolution is performed on the input data and a weights array, called the filter, to then produce a feature map. The filter actually slides over the input, and at every time step, a matrix multiplication is performed. This is repeated for each input parameter (feature) and the result is summed into a new feature map. In case of sequences or time series, often dilated causal convolutions are used.

Causal means that the output of the filter does not depend on future input time steps. Dilated means that the filter is applied over a range larger than its length by skipping input time steps with a certain step. By stacking dilated convolutions, the network can look further back into history (i.e., the receptive field) with just a few layers, while still preserving the input resolution throughout the network (i.e., the number of time steps in the sequence) as well as the computational efficiency. Often, each additional layer increases the dilation factor exponentially, as this allows the receptive field to grow exponentially with the network depth. This principle is shown in Figure 1 for a filter width of two time-steps.

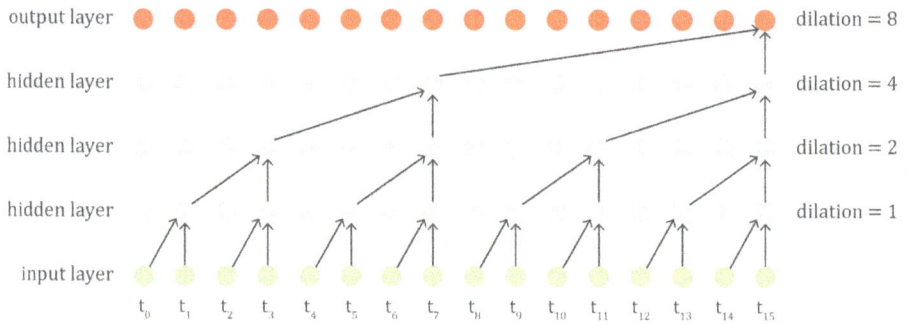

Figure 1. The causal dilated convolutions allow an output time step to receive information from a larger range of input time steps (i.e., the receptive field) with an increasing number of hidden layers. In the presented scheme, a filter width of two, four layers and one stack results in a receptive field of sixteen input steps.

The architecture of the CNN network used in this paper, shown in Figure 2 is based on the Wavenet architecture [16] and is developed using Keras 2.2.4 [17]. The network consists of stacked 'residual blocks', followed by two final convolutional layers. By layering multiple residual blocks, a larger receptive field is obtained. The dilation can be exponentially increased for a number of layers and then repeated, e.g., $2^0, 2^1, 2^2, \ldots, 2^9, 2^0, 2^1, 2^2, \ldots, 2^9, 2^0, 2^1, 2^2, \ldots, 2^9$, for filter width two. These repetitions of layered residual blocks are called stacks. The combination of the filter width, number of layers and number of stacks defines the length of the receptive field.

Each residual block contains three important elements that give the network its prediction strength: a gated activation unit, residual and skip connections and global conditioning. The gated activation unit starts with a causal dilated convolution, which then splits, passes through either a tanh or sigmoid activation and finally recombines via element-wise multiplication. The tanh activation branch can be interpreted as a learned filter and the sigmoid activation branch as a learned gate that regulates the information flow from the filter [18]. Recurrent neural networks such as the Long Short-Term Memory (LSTM) and Gated Recurrent Unit (GRU) use similar gating mechanisms to control the flow of information. The gated activation unit can mathematically be represented by Equation (1), where, W corresponds to the learned dilated causal convolution weights and f and g denote filter and gate, respectively:

$$z = \tanh\left(W_f * x\right) \odot \sigma\left(W_g * x\right) \tag{1}$$

The skip connections allow lower level signals to pass unfiltered to the final layers of the network. Hence, earlier feature layer outputs are preserved as the network passes forward signals for final prediction processing. This allows the network to identify different aspects of the time series, i.e., strong autoregressive components, sophisticated trend and seasonality components, as well as trajectories difficult to spot with the human eye. Residual connections allow each block's input to bypass the gated activation unit, and then add that input to the gated activation unit output. This helps allow for the possibility that the network learns an overall mapping that acts almost as an identity function, with the input passing through nearly unchanged. The effectiveness of residual connections is still not fully understood, but a compelling explanation is that they facilitate the use of deeper networks by allowing for more direct gradient flow in backpropagation [19].

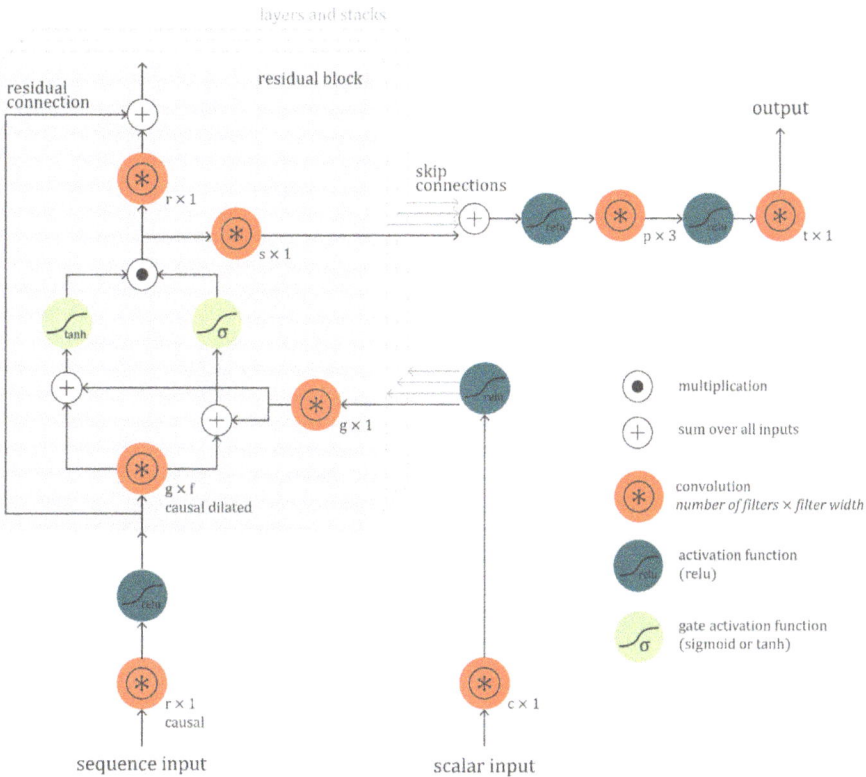

Figure 2. The used convolutional neural networks (CNN) architecture with residual blocks, skip connections and global conditioning, based on the Wavenet architecture.

Finally, global conditioning allows the network to produce output patterns for a specific context. For example, if different brick types are included, the network can be trained by feeding it the brick characteristics as additional input. In this case, the gated activation unit can mathematically be represented by Equation (2), where, V corresponds to the learned convolution weights and h is a tensor that contains the conditional scalar input and is broadcast over the time dimension:

$$z = \tanh\left(W_f * x + V_f * h\right) \odot \sigma\left(W_g * x + V_g * h\right) \tag{2}$$

2.2. Hyper-Parameter Optimization

In order to configure and train the network, the hyper-parameters of the network need to be set. For configuring the proposed architecture (Figure 2), these are:

- Filter width f of causal dilated convolution
- Number of c-filters for initial conditional connection
- Number of g-filters for gate connections
- Number of s-filters for skip connections

- Number of *r*-filters for residual connections
- Number of *p*-filters for the penultimate connection
- Number of layers of residual blocks
- Number of stacks of layered residual blocks

Additionally, there are hyper-parameters concerning the training process itself:

- Loss function
- Learning algorithm
- Learning rate
- Number of training epochs
- Batch size

The loss function is minimised during training by determining the neurons' optimal weights and is a measure of how good the network fits the data. In this optimisation, the root-mean-squared-error (RMSE) is used as the loss function, because it effectively penalises larger errors more severely. The learning algorithm defines how the neurons' weights are updated during the learning process. Many learning algorithms exist, but in this study, the Adam algorithm [20] is used, as the authors' previous experiments showed it to perform best for the current problem. The learning rate is the allowed amount of change to the neurons' weights during each step of the learning process. At extremes, a learning rate that is too large may result in too-large weight updates, causing the performance of the network to oscillate over training epochs. A too-small learning rate may never converge or may get stuck on a suboptimal solution. The learning rate must thus be carefully configured. The batch size is the number of training samples passed through the neural network in one step. The larger the batch size, the more memory is required during training. As the networks are trained on a computer with two NVIDIA RTX 2070 GPU's, each with 8 GB RAM, the available memory is limited. For this reason, the batch size is fixed to four samples. After each batch, the network's weights are updated. When all batches have passed through the network once, one training epoch is completed. The number of training epochs is the number of times the entire training dataset is passed through the neural network. The more often the network is exposed to the data, the better it becomes at learning to predict. However, too much exposure can lead to overfitting: the network's error on the training data is small but when new data is presented to the network, the error is large. This is prevented by stopping training if the error on the validation dataset no longer decreases, a mechanism called 'early stopping'.

To reduce the training time during the optimisation process, two measures are taken: Firstly, the training set contains only 256 samples, which reduces the number of batches in each epoch. Secondly, each neural network is trained for a maximum of only 50 epochs, and training is stopped earlier if the RMSE on the validation set (containing 64 samples) decreases less than 0.001 over 5 epochs. These measures reduce training time successfully, but do not allow for reaching the best prediction performance, as both the number of epochs and the number of samples in the training set are too small. However, this approach allows for identifying the hyper-parameter combinations that converge fastest and are thus likely to perform best. Table 1 gives an overview of all hyper-parameters that need to be fine-tuned, in order to get optimal prediction results. Because evaluating all possible combinations in a full factorial way would be extremely expensive, optimization of these hyper-parameters is done via the Grey-Wolf Optimiser (GWO) [12]. It is a population-based meta-heuristic based on the leadership hierarchy and hunting mechanism of grey wolves in nature. Grey wolves live in a pack in which alpha (α), beta (β), delta (δ) and omega (ω) wolves can be identified. Positioned on top of the pack, the α-wolf decides on the hunting process and other vital activities. The other wolves should follow the α-wolf's orders. The ß-wolves help the α-wolf in decision-making. The δ-wolves have to submit to the α- and ß-wolves, but dominate the ω-wolves, who are considered the scapegoats of the pack. In the GWO, the fittest solution is considered as α, and the second and third fittest solutions are named ß and δ, respectively. The rest of the solutions are ω. In search of the optimal solution, the α-, ß-, and δ-solutions guide the direction, and the ω-solutions follow. The three best solutions are saved and the other search agents (ω) are obligated to update their positions according to the positions of the best search agents.

Table 1. The search range of the hyper-parameters.

Hyper-Parameter	Range
Number of filters c	$(2^5; 2^9)$
Number of filters g	$(2^5; 2^9)$
Number of filters s	$(2^5; 2^9)$
Number of filters r	$(2^5; 2^9)$
Number of filters p	$(2^5; 2^9)$
Filter width f	$(2; 24)$
Number of layers	$(1; 8)$
Number of stacks	$(1; 4)$
Learning rate	$(0.0001; 0.01)$

In this study, 10 search agents are deployed to explore and exploit the search space over 100 iterations. If the best solution does not change for 25 iterations, the search algorithm is stopped. This is repeated for five independent runs as different runs might end with different optimal solutions. The RMSE on the validation set is used to evaluate the fitness of the solutions.

2.3. Performance Evaluation

Once the GWO algorithm has finished, the ten best solutions (lowest RMSE) of all runs are trained fully to reach the networks' full prediction potential, by using a training set of 768 samples, with a validation set of 192 samples. A maximum of 200 epochs is set, with early stopping if the RMSE decreases less than 0.001 over 20 epochs. Each combination is trained five times, to overcome initialisation differences. Note that the size of the training dataset is chosen rather arbitrarily: this is based on previous experiments, showing that training on 786 samples resulted in better prediction performance compared to 256 training samples (for identical hyper-parameters). These numbers might not be optimal, i.e., a larger dataset might result in even better prediction performance or vice versa, a smaller dataset might provide equally satisfying results.

The performance of these 10 fully-trained neural networks is evaluated using three performance indicators: the root mean-square error (RMSE), the mean absolute error (MAE), and the coefficient of determination (R2), quantified as follows:

$$RMSE = \sqrt{\frac{1}{T}\sum(y-\hat{y})^2} \quad MAE = \frac{1}{T}\sum|y-\hat{y}| \quad R2 = 1 - \frac{\sum(y-\hat{y})^2}{\sum(y-\overline{y})^2} \tag{3}$$

where, y is the true output, $\hat{y}$ is the predicted output, $\overline{y}$ is the mean of the true output and T is the total number of data points. Additionally, the models' training time is evaluated.

Finally, the best performing network, defined as the one with the lowest RMSE on the validation dataset (192 samples), is selected. Because performance on the validation dataset is incorporated into the network's hyper-parameter optimisation, this final network's performance is tested using an independent test set, containing 256 samples. This way, an unbiased performance evaluation is obtained. The performance indicators are calculated for each target separately, to identify which targets are more or less accurately predicted. Subsequently, the network's output is used to predict the damage risks. These results are evaluated using the same performance indicators as described above.

3. Application

3.1. Hygrothermal Simulation Object

The calculation object in this study is a 1D cross-section of a massive masonry wall. The masonry wall is simplified to an isotropic brick layer—no mortar joints are modelled [21]—and an interior plaster layer of 1.5 cm. Note that no construction details, such as corners or embedded beams, are modelled.

To explore the capabilities of the proposed convolutional neural network, all characteristics and boundary conditions that are expected to significantly influence the hygrothermal performance of the 1D wall are considered probabilistic (Table 2). Variability in climatic conditions is included by using different years of climate data of four Belgian cities [22]. Since the aim is to predict the expected future performance of the wall, future climate data is used. Variability in the wall conditions is incorporated via uniform distributions of the wall orientation, solar absorption and exposure to wind-driven rain. The wind-driven rain load is calculated by using the catch ratio, as described in Reference [23]. The catch ratio relates the wind-driven rain (WDR) intensity on a facade to the unobstructed horizontal rainfall intensity and is a function of the reference wind speed and the horizontal rainfall intensity for a given position on the building facade and wind direction. In this model, variability in wall position and potential shelters, trees or surrounding buildings are reckoned with by the exposure factor. Additionally, the transiency and variation of the wind speed is taken into account in the convective heat transfer coefficient, via Equation (4) (EN ISO 06946), where, $h_0 = 4 \text{ W/m}^2\text{K}$ and $k_e = 1$.

$$h_c = h_0 + k_s \cdot v_{wind}^{k_e} \qquad (4)$$

Table 2. Probabilistic input parameters and distributions.

Parameter	Value
Exterior climate	D (Gent; Gaasbeek; Oostende, St Hubert)
Exterior climate start year	D (2020; 2047)
Wall orientation (degrees from North)	U (0; 360)
Solar absorption (-)	U (0.4; 0.8)
Ext. heat transfer coefficient slope k_s (J/m^{3}K)	U (1; 8)
WDR exposure factor (-)	U (0; 2)
Brick wall thickness (m)	U (0.2; 0.5)
Brick material	D (Brick 1; Brick 2; Brick 3)
Interior humidity load [24]	D (load A; load B)

U (a, b): uniform distribution between a and b; D (a, b): discrete distribution between a and b.

The exterior moisture transfer coefficient is related to the exterior heat transfer coefficient through the Lewis relation. The interior climate is calculated according to EN 15026 [24] and variability in building use is included by using two different humidity loads. Finally, to explore the CNN's capabilities to the maximum, three different brick types as well as a uniform distribution of the wall thickness are included as well. The basic characteristics of the used brick types can be found in Table 3 and Figure 3, which clearly show the variations in the bricks' moisture properties.

Table 3. Brick type characteristics.

Parameter	Brick 1	Brick 2	Brick 3
Dry thermal conductivity (W/m^{2}K)	0.87	0.52	1.00
Dry vapour resistance factor (-)	139.52	13.25	19.00
Capillary absorption coefficient (kg/m^{2}s$^{0.5}$)	0.046	0.357	0.100
Capillary moisture content (m^3/m^3)	0.128	0.266	0.150
Saturation moisture content (m^3/m^3)	0.240	0.367	0.250

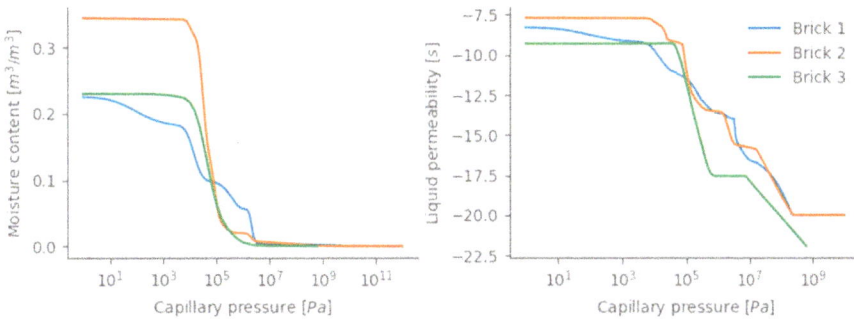

Figure 3. Brick type moisture properties; (left) the moisture retention curve and (right) the liquid permeability.

The remaining parameters are all variables either with small variations or of less importance for the current study of a 1D wall. Therefore, these boundary conditions are assumed deterministically. Table 4 gives an overview of the deterministic boundary conditions.

Table 4. Discrete input parameters.

Parameter	Value
Exterior surface	
Long wave emissivity	0.9
Interior surface	
Total heat transfer coefficient h (W/m^{2}K)	8
Moisture transfer coefficient β (s/m)	3×10^{-8}
Initial conditions	
Initial temperature (°C)	20
Initial relative humidity (%)	50

When evaluating the hygrothermal performance of a massive masonry wall, one is typically interested in frost damage at the exterior surface, decay of embedded wooden floors and mould growth on the interior surface [3,25–27]. The latter is mainly important in the case of thermal bridges and of less importance in 1D simulations. Table 5 gives an overview of frequently used prediction models for these damage patterns, and the required hygrothermal time series to evaluate them. Figure 4 schematically presents the two-dimensional (2D) building component (top) and the modelled 1D mesh (bottom) and indicates at which positions the hygrothermal performance is monitored. The simulations were performed using the hygrothermal simulation environment Delphin 5.8 [28], and a simulation period of four years was adopted. As most damage prediction models require hourly data, an hourly output frequency is used.

Table 5. Damage prediction models and required Delphin output.

Damage Pattern	Prediction Model	Required Hygrothermal Time Series
Frost damage	Moist freeze-thaw cycles	T, RH, saturation degree
Decay of wooden beam ends	VTT wood decay model	T, RH
Mould growth	Updated VTT mould growth model	T, RH

T: temperature; RH: relative humidity.

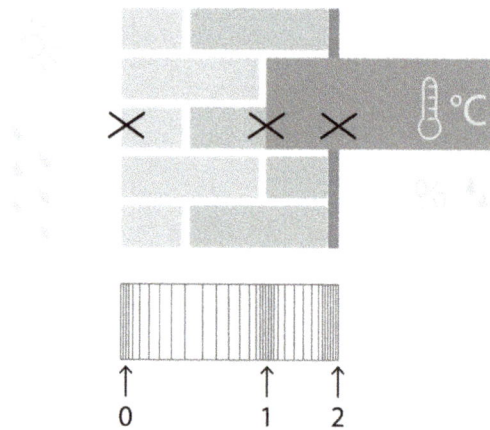

Figure 4. A schematic representation of the two-dimensional (2D) building component (top) and the modelled one-dimensional (1D) mesh (bottom), with indication at which positions the hygrothermal performance is monitored.

The frost damage risk is evaluated via the number of moist freeze-thaw cycles at 0.5 cm from the exterior surface. A 'moist' freeze-thaw cycle is a freeze-thaw cycle that occurs in combination with a moisture content high enough to induce frost damage [3]. In this study, the critical moisture content is defined as a moisture content above 25% of the saturated moisture content. Note that this is a rather arbitrary value, as currently no precise prediction criterion is at hand. An indication of the decay risk of wooden beam ends can be made using the VTT wood decay model, which calculates the percentage of mass loss of the wooden beam end based on the temperature and relative humidity [29]. Note, however, that in this 1D wall study, solely a rough indication of the wood decay risk is acquired, as two- and three-dimensional heat and moisture transport, as well as potential air rotations around the wooden beam end, are neglected [30]. At the interior surface, a too-high relative humidity entails a risk on mould growth. The mould growth risk can be estimated by the VTT mould growth model, which calculates the Mould Index based on the fluctuation of the temperature and relative humidity [31]. The Mould Index is a value between 0 and 6, going from no growth to heavy and tight mould growth. In the updated VTT model, the expected material sensitivity to mould growth is implemented as well. In this study, the materials are assumed to belong to the class 'very sensitive'.

3.2. Training the Convolutional Neural Network

The training and validation datasets are obtained by sampling the input parameters described above multiple times, using a Sobol sampling scheme [32], and simulating the deterministic HAM model once for each sampled input parameter combination. In this study, in total, 960 samples are used. The network is trained to predict hygrothermal time series as requested for the damage prediction models (see Table 5), based on the input in Tables 2 and 3. The inputs are pre-processed to facilitate learning. The scalar parameters 'wall orientation', 'exterior heat transfer coefficient slope', 'solar absorption' and 'rain exposure' are integrated in the exterior climate time series but also preserved as a separate scalar input parameter, to condition the network (see Figure 2). The categorical parameters 'start year' and 'interior humidity load' are incorporated into the climate time series. The categorical parameter 'brick type' is replaced by scalar parameters of the characteristics in Table 3. This simplifies the network architecture and allows more flexibility on using brick types with differing characteristics. This results in 6 input time series (exterior temperature, exterior relative humidity, wind-driven rain load, short-wave radiation, interior temperature and interior relative humidity) and 10 scalar inputs (exterior heat transfer coefficient slope (Equation (4)), rain exposure factor, solar absorption, wall orientation, brick wall thickness and the 5 brick characteristics from Table 3.

Before presenting the input and output data to the neural network, all data are standardised (zero mean, unit variance). This ensures that all features are on the same scale, which allows weighting all features equally in their representation. Standardising the output data ensures that errors are penalised equally for all targets.

4. Results and Discussion

4.1. Hyper-Parameter Optimization

The results of the hyper-parameter optimization show that some (combinations of) parameters have a significant influence on the network's performance, while others are less important. Figure 5 shows the RMSE on the validation dataset of all GWO candidate solutions, in function of the receptive field and the filter width of the causal dilated convolution. This figure clearly shows that a receptive field of at least 14 months (10,224 h) is required to obtain a low RMSE. The length of the receptive field is defined by the filter width, number of layers and number of stacks, and determines how many past-input time steps the network can use to predict the current output time step. Hence, it makes sense that the receptive field has a threshold below which the network does not perform well, as it cannot access enough information. Figure 5 also shows that a low RMSE can be obtained for all filter widths. Note that this is not the case for filter widths below five, as these require a large number of layers and stacks (cfr. receptive field), which caused out-of-memory errors on the used hardware. If more GPU memory is available, one might not run into this problem. Additionally, Figure 6 (top) shows that using multiple stacks results in a slightly lower RMSE, compared to only one stack. Adding extra stacks to the network allows for increasing the depth—and thus the complexity of the model—without increasing the receptive field exponentially. Indeed, if the receptive field becomes much larger than the actual time series length, the computational efficiency decreases. On the other hand, Figure 6 (top) also shows that the training time increases with the number of stacks. The number of layers has a similar influence on the training time (not shown here), but not on the prediction performance, provided that the receptive field is large enough. Hence, if the number of stacks were fixed, a large filter width would require fewer layers and thus shorter training time, compared to a smaller filter width, while both options would yield similar prediction performance.

Regarding the number of filters for the different connections, some tendencies can be observed in Figure 7 but there appears to be no clear relationship with the validation RMSE—with the exception that 32 filters is too few for all connections. However, the number of filters has a significant impact on the training time, as shown by Figure 6 (bottom). Combined with the filter width and the number of layers and stacks, the number of filters determines the number of trainable parameters. Hence, using fewer filters, for the same combination of filter width, layers and stacks, generally results in a lower training time.

Finally, Figure 8 indicates that the optimal learning rate for networks with larger filter widths is around 0.0015, whereas networks with smaller filters widths (i.e., deeper networks) seem to perform better with a larger learning rate around 0.003.

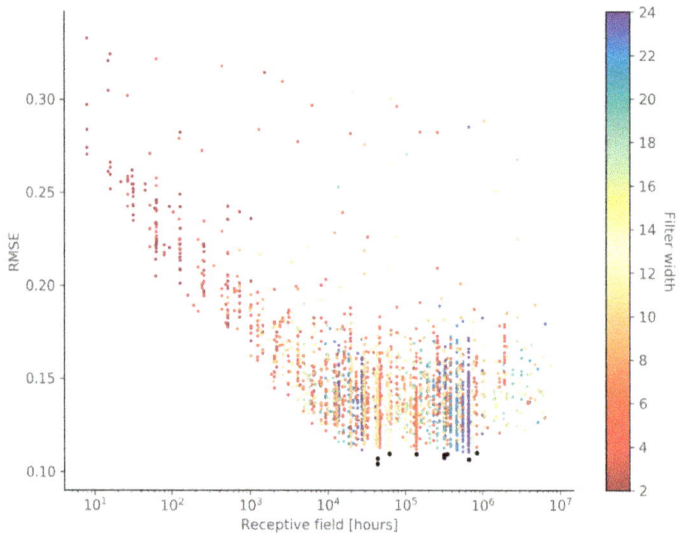

Figure 5. This scatterplot of all Grey-Wolf Optimiser (GWO) candidate solutions shows that a receptive field of at least 14 months (10,224 h) is required to obtain a low root mean-square error (RMSE) on the validation set, and that a low RMSE can be obtained for all filters widths, once above this receptive field threshold. The 10 best solutions are indicated in black.

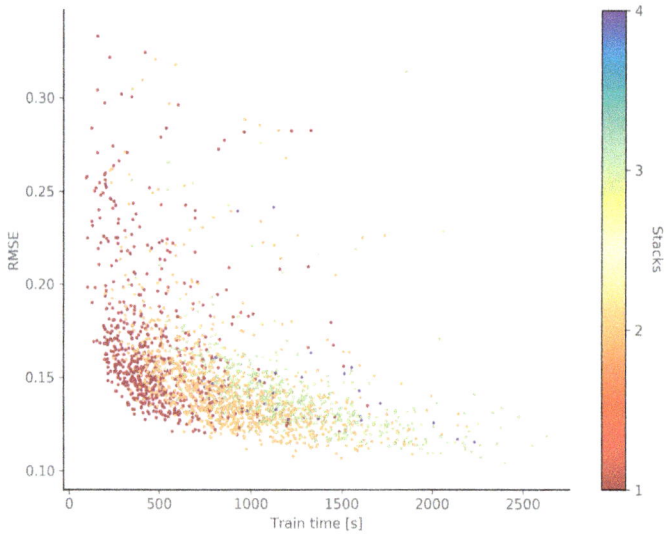

Figure 6. *Cont.*

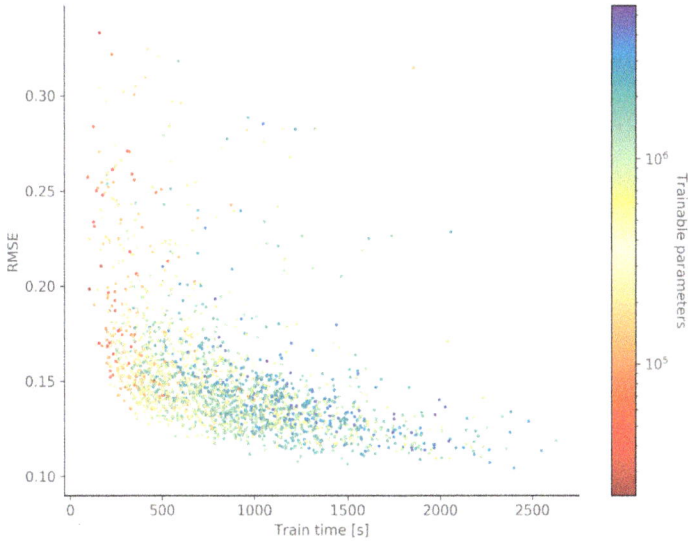

Figure 6. This scatterplot of all GWO candidate solutions shows a clear relation between the performance (RMSE), the training time and the number of stacks (top) or the number of parameters (determined by the number of layers, stacks and filters) (bottom).

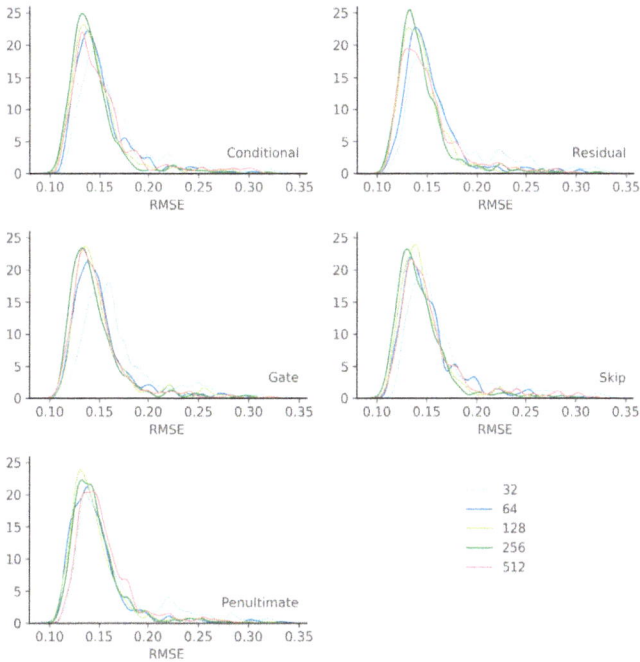

Figure 7. This distribution plot of all GWO candidate solutions shows some tendencies but no clear relation between the number of filters and the RMSE on the validation set.

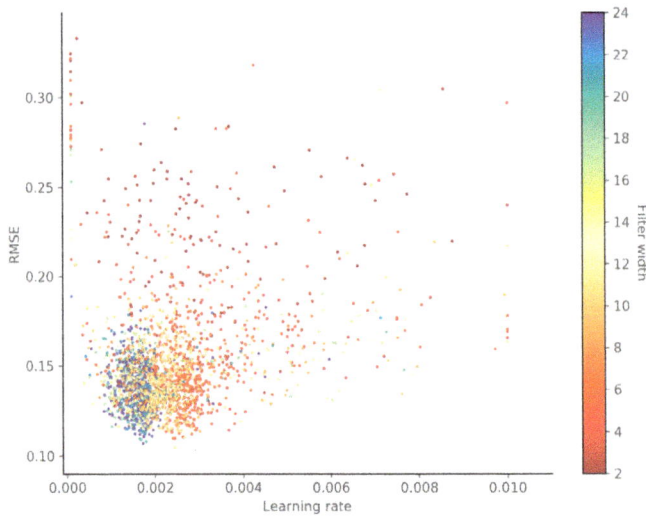

Figure 8. This scatterplot of all GWO candidate solutions shows that the optimal learning rate depends on the used filter width: larger filter widths require smaller learning rates.

4.2. Performance Evaluation

The ten best solutions of the Grey-Wolf Optimiser algorithm are shown in Table 6. and confirm the overall findings described above. The performance indicators of training five repetitions of these ten combinations on the entire training dataset are shown in Figure 9. The results indicate that no single hyper-parameter combination performs significantly better than the others. As long as the receptive field is large enough (>14 months), the network is deep enough (≥2 stacks) and the learning rate is in the range (0.0015; 0.003), the other hyper-parameters appear to have only a minor influence on the prediction performance. Furthermore, due to weight and bias initialisation differences, training a network with identical hyper-parameters twice does not necessarily result in the same prediction performance, as can be observed in Figure 9. Hence, it is best to repeat training a few times, and select the best performing network afterwards. Figure 9 also confirms that, in general, deeper networks with more layers are slower to train. Finally, note that increasing the maximum number of training epochs and the size of the training dataset resulted in a much lower RMSE, compared to the results obtained by the GWO algorithm. This underlines the importance of a representative training dataset, as well as allowing enough training iterations.

Table 6. The ten best performing solutions of the Grey Wolf Optimiser algorithm.

	Conditional Filters	Gate Filters	Skip Filters	Residual Filters	Penultimate Filters	Filter Width	Layers	Stacks	Learning Rate
1	256	512	256	128	64	11	3	3	0.00245
2	256	256	256	512	64	24	3	2	0.00172
3	256	512	256	128	128	11	3	3	0.00220
4	256	256	256	256	128	20	3	2	0.00179
5	128	512	512	256	128	20	3	2	0.00167
6	32	512	256	256	64	20	3	2	0.00164
7	128	64	128	256	256	6	5	3	0.00245
8	128	64	128	256	256	7	5	3	0.00241
9	128	512	128	64	128	12	3	3	0.00266
10	128	128	128	128	128	6	6	3	0.00263

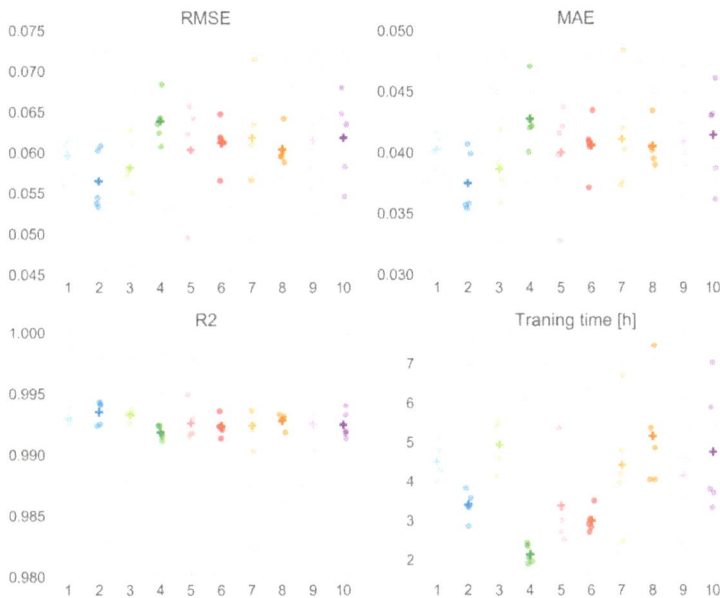

Figure 9. The performance indicators and training time for the ten best-performing hyper-parameter combinations, after being trained fully. For each combination, the dots indicate each repetition's result, whereas the cross indicates the average over all five repetitions.

Model 2 performs best on average, but one repetition of model 5 performs better than all other models. Hence, this model's performance is evaluated using the independent test set. Figure 10 shows an example of a test set sample prediction. The performance indicators shown above each panel are calculated on the standardised output for each target separately, as this indicates the difference in accuracy between targets. It is clear that the chosen network is able to predict all hygrothermal outputs quite accurately. These hygrothermal predictions can be used to evaluate damage risks, as described in Section 3.1. Figure 11 shows the damage predictions (orange) using the networks' output (for the sample shown in Figure 10), which are in almost perfect agreement with the damage predictions from the Delphin simulations (blue). By expressing the damage risks as single values, it is possible to show the damage prediction accuracy of all individual samples (Figure 12, top) and the cumulated distributions (Figure 12, bottom). The latter is a common presentation in a probabilistic assessment, as it gives information on the distribution of the expected performance taking into account all uncertainties. The frost damage risk is given by the total number of moist freeze-thaw cycles at the end of the simulated period, the mould growth risk on the interior surface is expressed as the maximum mould index over the whole simulated period and the wood decay risk is expressed as the total wood mass loss at the end of the simulated period. Figure 12 shows that the damage risk prediction, based on the networks' hygrothermal predictions, is quite accurate for most samples of the test set. The number of moist freeze-thaw cycles tends to be overestimated, due to small prediction errors in the temperature and relative humidity. A slightly lower temperature and/or relative humidity at one time step can lead to counting more freeze-thaw cycles compared to the true value. However, both the original hygrothermal model and the neural network predict a low number of moist freeze-thaw cycles, and a difference of a few cycles will likely not much influence the extent of the expected damage. In case of the mould index at the interior surface, the deviations are so small they are negligible. The wood decay risk tends to be slightly underestimated, but the overall agreement in cumulative distribution function is very good.

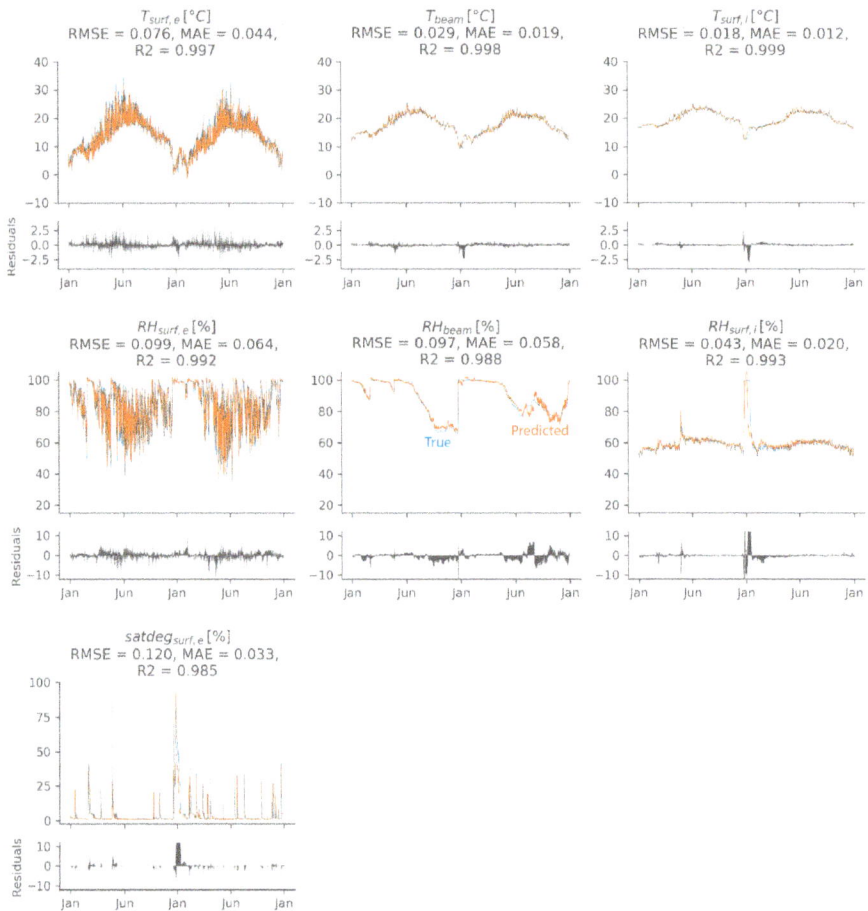

Figure 10. The network's hygrothermal predictions (orange) of a test set sample with high accuracy. The true value is shown in blue the prediction error is indicated in grey.

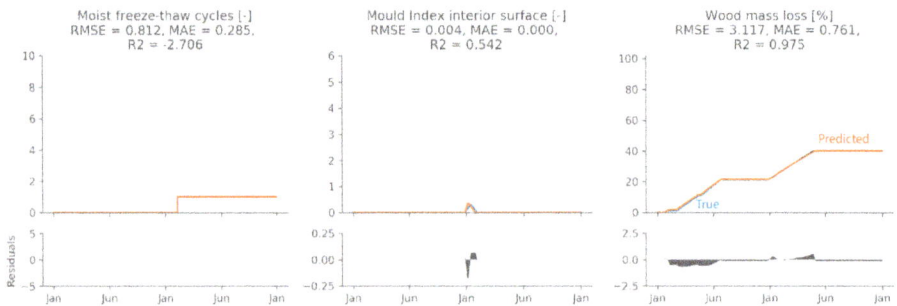

Figure 11. The damage predictions (orange) for the sample from Figure 10, using the network's hygrothermal predictions. The true value is shown in blue the prediction error is indicated in grey.

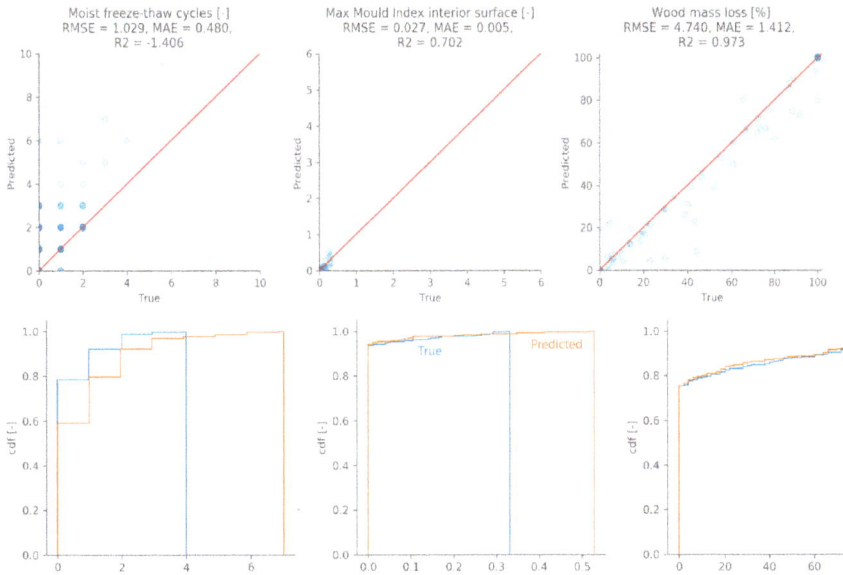

Figure 12. A comparison between the single-value damage predictions, as obtained using the network's predictions and the true damage predictions. The top panel shows the goodness-of-fit for the individual samples (transparency is used to indicate overlapping samples), the bottom panels show the cumulated distribution functions.

5. Conclusions

In this paper, convolutional neural networks were used to replace HAM models, aiming to predict the hygrothermal time series (e.g., temperature, relative humidity, moisture content). A strategy was presented to optimise the networks' hyper-parameters, using the Grey-Wolf Optimiser algorithm and a limited training dataset. This approach was applied to the hygrothermal response of a massive masonry wall, for which the prediction performance and the training time were evaluated. Based on the GWO optimisation, it was found that the receptive field—defined by the filter width, number of layers and number of stacks—has a significant impact on the prediction performance. For the current case study of massive masonry exposed to driving rain, it needs to span at least 14 months. The results also showed that good performance can be obtained for all filter widths, as long as the receptive field is large enough. Additionally, using multiple stacks resulted in slightly better performance compared to a single stack, as this allows adding complexity to the model, but also resulted in longer training time. The number of layers, determined by the filter width and the number of stacks to obtain a large enough receptive field, had a similar influence on the training time, but not on the prediction performance. Hence, if the number of stacks were fixed, a large filter width would require fewer layers and thus shorter training time, compared to a smaller filter width, while both options would yield similar prediction performance. The same applies to the number of filters for the different convolutional connections: the more filters that are used, the longer the training time becomes, without obvious benefit to the prediction performance. Finally, the learning rate was found to be optimal between 0.015 and 0.03, but only had a minor influence on prediction performance.

The 10 best-performing hyper-parameter combinations were trained further on a larger dataset. Of these, the best performing network was chosen and evaluated on an independent test set. These results showed that the proposed convolutional neural network is able to capture the complex patterns of the hygrothermal response accurately. To end, the predicted hygrothermal time series were used to calculate damage prediction risks, which were found to correspond well with the true damage

Energies **2019**, *12*, 3966

prediction risks. Hence, in conclusion, the proposed convolutional neural networks are very suited to replace time-consuming, standard HAM models.

Author Contributions: A.T. designed the model and the computational framework, performed the calculations and analysed the data, under supervision of H.J. and S.R. All authors discussed the results. A.T. wrote the manuscript with input from all authors.

Funding: This research was funded by the European Union's Horizon 2020 research and innovation program under grant agreement No 637268.

Conflicts of Interest: The authors declare no conflict of interest. The funders had no role in the design of the study; in the collection, analyses, or interpretation of data; in the writing of the manuscript, or in the decision to publish the results.

References

1. Van Gelder, L.; Janssen, H.; Roels, S. Probabilistic design and analysis of building performances: Methodology and application example. *Energy Build.* **2014**, *79*, 202–211. [CrossRef]
2. Janssen, H.; Roels, S.; Van Gelder, L. Annex 55 Reliability of Energy Efficient Building Retrofitting—Probability Assessment of Performance & Cost (RAP-RETRO). *Energy Build.* **2017**, *155*, 166–171.
3. Vereecken, E.; Van Gelder, L.; Janssen, H.; Roels, S. Interior insulation for wall retrofitting—A probabilistic analysis of energy savings and hygrothermal risks. *Energy Build.* **2015**, *89*, 231–244. [CrossRef]
4. Arumägi, E.; Pihlak, M.; Kalamees, T. Reliability of Interior Thermal Insulation as a Retrofit Measure in Historic Wooden Apartment Buildings in Cold Climate. *Energy Procedia* **2015**, *78*, 871–876. [CrossRef]
5. Gradeci, K.; Labonnote, N.; Time, B.; Köhler, J. A proposed probabilistic-based design methodology for predicting mould occurrence in timber façades. In Proceedings of the World Conference on Timber Engineering, Vienna, Austria, 22–25 August 2016.
6. Zhao, J.; Plagge, R.; Nicolai, A.; Grunewald, J. Stochastic study of hygrothermal performance of a wall assembly—The influence of material properties and boundary coefficients. *HVACR Res.* **2011**, *9669*, 37–41.
7. Janssen, H. Monte-Carlo based uncertainty analysis: Sampling efficiency and sampling convergence. *Reliab. Eng. Syst. Saf.* **2013**, *109*, 123–132. [CrossRef]
8. Van Gelder, L.; Janssen, H.; Roels, S. Metamodelling in robust low-energy dwelling design. In Proceedings of the 2nd Central European Symposium on Building Physics, Vienna, Austria, 9–11 September 2013; pp. 9–11.
9. Gossard, D.; Lartigue, B.; Thellier, F. Multi-objective optimization of a building envelope for thermal performance using genetic algorithms and artificial neural network. *Energy Build.* **2013**, *67*, 253–260. [CrossRef]
10. Bienvenido-huertas, D.; Moyano, J.; Rodríguez-jiménez, C.E.; Marín, D. Applying an arti fi cial neural network to assess thermal transmittance in walls by means of the thermometric method. *Appl. Energy* **2019**, *233–234*, 1–14. [CrossRef]
11. Tijskens, A.; Roels, S.; Janssen, H. Neural networks for metamodelling the hygrothermal behaviour of building components. *Build. Environ.* **2019**, *162*, 106282. [CrossRef]
12. Mirjalili, S.; Mirjalili, S.M.; Lewis, A. Grey Wolf Optimizer. *Adv. Eng. Softw.* **2014**, *69*, 46–61. [CrossRef]
13. Vereecken, E.; Roels, S.; Janssen, H. Inverse hygric property determination based on dynamic measurements and swarm-intelligence optimisers. *Build. Environ.* **2018**, *131*, 184–196. [CrossRef]
14. Borovykh, A.; Bohte, S.; Oosterlee, C.W. Conditional Time Series Forecasting with Convolutional Neural Networks. *arXiv* **2017**, arXiv:1703.04691.
15. Bai, S.; Kolter, J.Z.; Koltun, V. An Empirical Evaluation of Generic Convolutional and Recurrent Networks for Sequence Modeling. *arXiv* **2018**, arXiv:1803.01271.
16. van den Oord, A.; Dieleman, S.; Zen, H.; Simonyan, K.; Vinyals, O.; Graves, A.; Kalchbrenner, N.; Senior, A.; Kavukcuoglu, K. WaveNet: A Generative Model for Raw Audio. *arXiv* **2016**, arXiv:1609.03499.
17. Chollet, F. Keras. Available online: https://github.com/keras-team/keras (accessed on 15 May 2019).
18. van den Oord, A.; Kalchbrenner, N.; Vinyals, O.; Espeholt, L.; Graves, A.; Kavukcuoglu, K. Conditional Image Generation with PixelCNN Decoders. In Proceedings of the Conference on Neural Information Processing Systems, Barcelona, Spain, 5–10 December 2016.

19. He, K.; Zhang, X.; Ren, S.; Sun, J. Deep Residual Learning for Image Recognition. In Proceedings of the IEEE Conference on Computer Vision and Pattern Recognition (CVPR), Boston, MA, USA, 8–10 June 2015; Volume 19, pp. 107–117.

20. Kingma, D.P.; Ba, J. Adam: A Method for Stochastic Optimization. *arXiv* **2014**, arXiv:1412.6980.

21. Vereecken, E.; Roels, S. Hygric performance of a massive masonry wall: How do the mortar joints influence the moisture flux? *Constr. Build. Mater.* **2013**, *41*, 697–707. [CrossRef]

22. European Commission. *Climate for Culture: Damage Risk Assessment, Economic Impact and Mitigation Strategies for Sustainable Preservation of Cultural Heritage in Times of Climate Change*; European Commission: Brussels, Belgium, 2014.

23. Blocken, B.; Carmeliet, J. Spatial and temporal distribution of driving rain on a low-rise building. *Wind Struct. Int. J.* **2002**, *5*, 441–462. [CrossRef]

24. European Committee for Standardisation. *EN 15026:2007—Hygrothermal Performance of Building Components and Building Elements—Assessment of Moisture Transfer by Numerical Simulation*; European committee for Standardisation: Brussels, Belgium, 2007.

25. Harrestrup, M.; Svendsen, S. Internal insulation applied in heritage multi-storey buildings with wooden beams embedded in solid masonry brick façades. *Build. Environ.* **2016**, *99*, 59–72. [CrossRef]

26. Zhou, X.; Derome, D.; Carmeliet, J. Hygrothermal modeling and evaluation of freeze-thaw damage risk of masonry walls retrofitted with internal insulation. *Build. Environ.* **2017**, *125*, 285–298. [CrossRef]

27. Marincioni, V.; Marra, G.; Altamirano-medina, H. Development of predictive models for the probabilistic moisture risk assessment of internal wall insulation. *Build. Environ.* **2018**, *137*, 257–267. [CrossRef]

28. Delphin 5.8 [Computer Software]. Available online: www.http://bauklimatik-dresden.de/delphin (accessed on 15 May 2019).

29. Viitanen, H.; Toratti, T.; Makkonen, L.; Peuhkuri, R.; Ojanen, T.; Ruokolainen, L.; Räisänen, J. Towards modelling of decay risk of wooden materials. *Eur. J. Wood Wood Prod.* **2010**, *68*, 303–313. [CrossRef]

30. Vereecken, E.; Roels, S. Wooden beam ends in combination with interior insulation: An experimental study on the impact of convective moisture transport. *Build. Environ.* **2019**, *148*, 524–534. [CrossRef]

31. Ojanen, T.; Viitanen, H.; Peuhkuri, R.; Lähdesmäki, K.; Vinha, J.; Salminen, K. Mold Growth Modeling of Building Structures Using Sensitivity Classes of Materials. In Proceedings of the Thermal Performance of the Exterior Envelopes of Buildings XI, Clearwater Beach, FL, USA, 5–9 December 2010; pp. 1–10.

32. Hou, T.; Nuyens, D.; Roels, S.; Janssen, H. Quasi-Monte-Carlo-based probabilistic assessment of wall heat loss. *Energy Procedia* **2017**, *132*, 705–710. [CrossRef]

energies

MDPI

Article

Self-Heating Ability of Geopolymers Enhanced by Carbon Black Admixtures at Different Voltage Loads

Lukáš Fiala [1,*], Michaela Petříková [1], Wei-Ting Lin [2], Luboš Podolka [3] and Robert Černý [1]

1   Department of Materials Engineering and Chemistry, Faculty of Civil Engineering, Czech Technical University in Prague, Thákurova 7, 166 29 Prague 6, Czech Republic; michaela.petrikova@fsv.cvut.cz (M.P.); cernyr@fsv.cvut.cz (R.C.)
2   Department of Civil Engineering, College of Engineering, National Ilan University, No.1, Sec. 1, Shennong Rd., I-Lan 260, Taiwan; wtlin@niu.edu.tw
3   Department of Civil Engineering, Faculty of Technology, Institute of Technology and Business in České Budějovice, Okružní 517/10, 370 01 České Budějovice, Czech Republic; podolka@mail.vstecb.cz
*   Correspondence: fialal@fsv.cvut.cz; Tel.: +420-22435-7125

Received: 20 September 2019; Accepted: 25 October 2019; Published: 29 October 2019

Abstract: Sustainable development in the construction industry can be achieved by the design of multifunctional materials with good mechanical properties, durability, and reasonable environmental impacts. New functional properties, such as self-sensing, self-heating, or energy harvesting, are crucially dependent on electrical properties, which are very poor for common building materials. Therefore, various electrically conductive admixtures are used to enhance their electrical properties. Geopolymers based on waste or byproduct precursors are promising materials that can gain new functional properties by adding a reasonable amount of electrically conductive admixtures. The main aim of this paper lies in the design of multifunctional geopolymers with self-heating abilities. Designed geopolymer mortars based on blast-furnace slag activated by water glass and 6 dosages of carbon black (CB) admixture up to 2.25 wt. % were studied in terms of basic physical, mechanical, thermal, and electrical properties (DC). The self-heating ability of the designed mortars was experimentally determined at 40 and 100 V loads. The percolation threshold for self-heating was observed at 1.5 wt. % of carbon black with an increasing self-heating performance for higher CB dosages. The highest power of 26 W and the highest temperature increase of about 110 °C were observed for geopolymers with 2.25 wt. % of carbon black admixture at 100 V.

Keywords: geopolymers; ground-granulated blast-furnace slag; carbon black; self-heating

---

1. Introduction

Building materials with new functional properties that extend their usability in sophisticated applications, so-called multifunctional or smart materials, are currently in high demand by the construction industry. Studies dealing with their design, experimental determination of material properties, and testing of newly achieved abilities have been, and still are, mainly focused on cementitious composites. A comprehensive review dealing with a definition and classification of smart concretes and structures and possible applications introduced by Han et al. [1] showed that a variety of possible enhancements exist. Some of the new functional properties, such as self-sensing, self-heating, energy harvesting, or electromagnetic shielding/absorbing, are crucially dependent on electrical properties that are, in the case of common aluminosilicate-based building materials, often very poor. Therefore, some electrically conductive admixtures are necessary for the formation of a conductive net within the material matrix. Much effort has been devoted to studies dealing with influence of carbon-based and metallic admixtures on new functional properties of cementitious materials. For example, Rana et al. [2] introduced a review focused on utilization of carbon-based materials such as carbon fibers (CFs), carbon nanofibers (CNFs), and carbon nanotubes (CNTs) in self-sensing cementitious materials; Han et al. [3] reviewed the self-sensing properties of cementitious materials with nanocarbon admixtures, namely CNFs, CNTs, and nano graphite platelets (NGPs); the review of Li et al. [4] was focused on research performed on cementitious composites with nano titanium dioxide (NT) and concluded that such materials possess self-sensing properties; and Pisello et al. [5] performed a detailed characterization of cementitious materials with multiwalled carbon nanotubes (MWCNTs), CNFs, carbon black (CB), and graphene nanoplatelets (GNPs) and, based on results, concluded that MWCNTs optimized piezoresistive properties and the tested nanofillers could be useful for cementitious smart materials and energy efficiency optimization.

The self-heating ability of cement-matrix and polymer-matrix composites with steel fibers (SFs) and CFs for deicing and space heating was reviewed by Chung [6]. Gomis et al. [7] studied in experimental and theoretical ways the self-heating ability of cement pastes with graphite powder (GP), carbon fiber powder (CFP), CFs, CNFs, and CNTs and, according to experiments conducted on samples with dimensions $100 \times 100 \times 10 \text{ mm}^3$ loaded by 50, 100, and 150 V DC, concluded that the self-heating ability of such materials is convenient in preventing the formation of ice layers in transportation infrastructures. Armoosh and Oltulu [8,9] investigated the self-heating ability of cementitious composites with metallic admixtures, iron, copper, and brass shavings, up to 20%, and proved the self-heating ability under a voltage load in the range of 20–60 V.

Wei et al. presented within several works the energy harvesting ability of cementitious materials with different electrically conductive admixtures, namely expanded graphite (EG) [10], CNT [11], and CF [12]. Despite the fact that the energy-harvesting efficiency of such materials is not high, it is promising large surface area of constructions usable for harvesting securing reasonable energy profit.

In general, cementitious materials are currently the most frequently used construction materials worldwide. Global production of cement has grown rapidly in recent years [13,14], and it is the third-largest source of anthropogenic emissions of carbon dioxide, after fossil fuels, deforestation, and other land-use changes. Global cement production has increased more than 30 times since 1950, and global process emissions in 2017 were 1.48 ± 0.20 Gt CO_2. Cumulative emissions from 1928 to 2017 were 36.9 ± 2.3 Gt CO_2, of which 70% has occurred since 1990 [15]. Taking into account the high negative impact of cement production on the environment, the design of building materials with comparable material properties to cementitious materials, but with a lower environmental impact of their production, is justified.

Geopolymers are inorganic materials of an environmentally friendly nature thanks to their fundamental component, so-called precursor, that is usually waste or byproduct originating from various types of industrial production [16]. A comprehensive review of precursors and alkali activators was introduced by Ma et al. [17], with a conclusion that geopolymers present better mechanical properties, a higher durability, and a more desirable structural performance compared to their conventional counterparts. Based on experiments, Albitar et al. [18] concluded that geopolymers are more chemically stable, superior to conventional concrete in an acidic environment, and exhibit lower deterioration of mechanical properties under chemical attacks. Good resistance at high temperatures was experimentally proved by Zuda et al. [19]. Taking into account the good material properties and lower impact on the environment than of cementitious materials, geopolymers can find an application in building practices.

Geopolymer binders are formed by the reaction of alkalis with amorphous aluminosilicate-rich precursors whose composition determines the material structure that is formed during hydration. In high-calcium systems (blast-furnace slag), typically calcium alumina-silicate hydrated gel (C-A-S-H) is formed [20], whereas in low-calcium systems (fly-ash, metakaolin, clay), sodium alumina-silicate hydrated gel (N-A-S-H) is present [21]. Alkali activation can be carried out by various alkali-activators, such as alkali hydroxides, weak or strong acid salts, silicates, aluminates, or aluminosilicates [22]. In general, the higher the alkalinity of the activator, the faster the initial reaction of the activator with precursor. The differences in heat evolution of slag activated by NaOH, water glass, and a combination of NaOH and water glass were presented by Altan and Erdogan [23], and a faster initial reaction was observed in NaOH by isothermal calorimetric measurements by Haha et al. [24]. Mostly used alkaline activators are mixtures of sodium or potassium hydroxide (NaOH, KOH) with sodium or potassium water glass (n·SiO$_2$·Na$_2$O, n·SiO$_2$·K$_2$O) [25].

Multifunctional geopolymers can be designed in a similar way to multifunctional cementitious materials by the addition of electrically conductive admixtures [26]. However, the design of multifunctional geopolymers and their acquired abilities are not so well explored. Rovnaník et al. [27] compared self-sensing properties of alkali-activated slag mortars with Portland cement mortars and concluded that, due to some content of iron particles in slag, an applicable sensitivity is evident in practice, even without any electrically conductive admixture, whereas in the case of Portland cement, mortar self-sensing is detectable but not sufficient for practical applications. Another similar study performed by Rovnaník et al. [28] dealt with the self-sensing ability of a geopolymer mortar based on slag activated by water glass with GP admixture. They concluded that such materials exhibit a self-sensing ability but with a significant decrease in compressive strength. Concerning the self-heating ability, it was experimentally confirmed on small alkali-activated slag samples with a CB admixture in the amount of 8.89 wt. % at 32.1 and 41.5 V by Fiala et al. [29]

Slag as a high-calcium precursor is a solid waste generated by the iron and steel industry. In 2014, slag was produced in the amount of 250 Mt within the 1.6 Gt of global steel production [30]. In 2013, the annual slag production of one of the leading producers, China, reached more than 100 million tons with just 29.5% utilization rate, which is very low in comparison to industrial countries. The utilization rate reaches 98.4% in Japan, 87.0% in Europe, and 84.4% in the United States. As of 2016, more than 300 million tons of accumulated steel slag has not been used effectively in China, which, taking into account large steel slag emissions, causes an important environmental problem for China [31]. Slag in granulated form is a precursor that can be relatively easily alkali-activated, and originating geopolymers can be used in the construction industry.

Within the research presented in this paper, granulated blast-furnace slag (GBFS) was used as a precursor for alkali activation by water glass, and CB admixtures were added in various dosages in order to enhance the effective electrical properties of the designed geopolymers that would be promising in terms of the self-heating ability. Subsequently, material properties involving basic physical, mechanical, thermal, and electrical properties were experimentally determined, and self-heating tests were conducted in order to verify the self-heating ability of such materials. It was observed that the self-heating ability of the tested materials started from a CB amount of 1.5 wt. %, and such material is able to generate heat at a DC voltage of 40 V leading to a small temperature increase. The best self-heating performance was observed for geopolymers with 2.25 wt. % of CB at 100 V, where the temperature increase was about 110 °C in approximately one hour.

2. Materials and Methods

A high-calcium precursor, GBFS SMŠ 380 (39.8% CaO), produced by Kotouč Štramberk Ltd. was activated by water glass Susil produced by Vodní sklo a.s. GBFS is an industrial byproduct of iron production with a fineness of 380 m^2·kg^{-1} (Blaine). The average grain size of the slag particles determined by laser granulometry was d_{50} = 15.5 µm and d_{90} = 38.3 µm. Activation was performed by sodium silicate solution (water glass Susil MP 2.0 with a molar ratio SiO$_2$/Na$_2$O = 2.07). Three normalized CEN fractions of fine quartz sand (PG1, PG2, PG3) produced by Filtrační písky, Ltd., that complied with the ČSN EN 196-1 standard were used as a filler. The effective electrical properties of the composites were enhanced by CB VULCAN 7H. CB is essentially elemental carbon in the form of spherical particles and aggregated clusters of those particles with a high surface area and high electrical conductivity. The average grain size of CB particles was d_{50} = 0.52 µm and d_{90} = 17.6 µm. In Figure 1, the particle size distribution of CB VULCAN 7H determined by laser granulometry is presented.

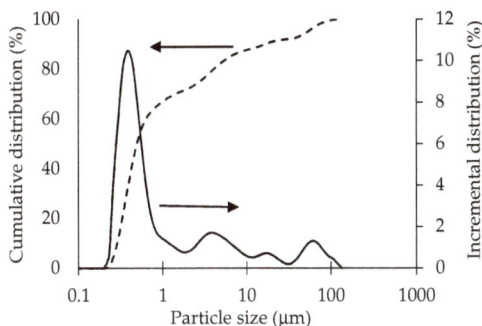

Figure 1. Particle size distribution of CB VULCAN 7H determined by laser granulometry.

CB VULCAN 7H is mainly manufactured for tires and industrial rubber production [32]. Its production is performed by thermal cracking of heavy aromatic feedstock, such as oil, in a hot flame. Oil is injected into a furnace hot flame zone where hydrocarbons are cracked to carbon and hydrogen by means of quenching the flame by water. Because of the reasonably high electrical conductivity and high surface area of such a way of processing CB, it can be used to optimize the electrical properties of polymers [33] and inorganic building materials [34,35].

In Table 1, the compositions of geopolymers with an optimized amount of mixing water are given. Seven different mixtures, the reference geopolymer (CB 0), and geopolymers with CB in the amounts of 0.75 wt. %, 1.25 wt. %, 1.5 wt. %, 1.75 wt. %, 2 wt. %, and 2.25 wt. % were designed and prepared by the following procedure. First, suspensions (10% and 15%) were prepared by adding a given amount of CB powder into water with nonionic surfactant Triton X-100 (Sigma-Aldrich, St. Louis, MO, USA) in the form of 0.5% solution and stirred by homogenizer IKA ULTRA-TURRAX for 15 min. An initial amount of additional water was then added to the GBFS suspension and stirred by a mixer for several minutes. In order to eliminate foaming during mixing leading to the formation of large pores, 1% solution of siloxane-based air-detraining agent Lukosan S (Lučební závody, Kolín, Czech Republic) was added. Subsequently, three fractions of sand were added to the mixture and stirred again by a mixer for several minutes. Consistency of fresh mixtures was tested according to the ČSN EN 1015-3 standard Determination of Consistence of Fresh Mortar by Flow Table, which is mainly used for cementitious mortars but is used also for alkali-activated ones [36]. The water-to-slag ratio of the mixtures was adjusted so that the average base diameter was equal to 160 mm, which is within the plastic range (140–200 mm) closer to the dry consistency bounds. Final mixtures with optimized water-to-slag ratios were then poured into molds and covered by a plastic cover. After one day, solidified samples were demolded and placed into a water bath for 28 d. Before the experiments (except the experimental determination of mechanical properties), samples were dried in an oven and subsequently cooled down in desiccator with silica gel.

Table 1. Compositions of the studied geopolymers.

	Carbon Black (CB) 0	CB 0.75	CB 1.25	CB 1.5	CB 1.75	CB 2	CB 2.25
Granulated blast-furnace slag (GBFS) (g)	100	100	100	100	100	100	100
Water glass (g)	20	20	20	20	20	20	20
Sand PG1 (g)	100	100	100	100	100	100	100
Sand PG2 (g)	100	100	100	100	100	100	100
Sand PG3 (g)	100	100	100	100	100	100	100
CB suspension (%)	0	10	10	10	15	15	15
CB amount (g)	0	3	5	6	7	8	9
Water-to-slag ratio (−)	0.44	0.60	0.69	0.74	0.77	0.81	0.84

The bulk density was determined on dry samples with dimensions of $50 \times 50 \times 50$ mm^3 by means of the gravimetric method. The matrix density was determined on samples with dimensions of $50 \times 50 \times 50$ mm^3 by means of the vacuum saturation method. With respect to the bulk density and the matrix density, the total open porosity Ψ (%) was calculated by the following equation

$$\Psi = 100 \cdot \left(1 - \frac{\rho_v}{\rho_{mat}}\right), \qquad (1)$$

where ρ_v (kg·m^{-3}) is the bulk density, and ρ_{mat} (kg·m^{-3}) is the matrix density.

Mechanical properties were determined on three samples with dimensions of $40 \times 40 \times 160$ mm^3 according to the Czech Standard ČSN EN 196-1. Samples cured for 28 d were first tested in terms of flexural strength by a three-point bending test. The length between the supports was 100 mm, and the loading rate was 0.15 mm/min. The compressive strength was then determined on six halves of the prisms originating from the previous flexural strength tests.

Thermal properties were determined on samples with dimensions $70 \times 70 \times 70$ mm^3 by a commercial device ISOMET 2114 (Applied Precision, Ltd.) attached by a surface probe by means of the transient heat-pulse method. Such measurements were based on an analysis of the temperature response to the generated heat flow pulses. Heat flow was induced by electric heating using a resistor heater placed in the probe having direct thermal contact with the surface of the sample. First, the probe was heated up, and, subsequently, temperature decrease was monitored after the heater was turned off. With the known geometry of the probe and a decrease of the temperature due to dissipation of the heat in the sample, thermal properties were identified.

Electrical properties were measured in a two-probe arrangement in DC regime. First, the samples with dimensions $50 \times 50 \times 50$ mm^3 were attached to electrodes where two opposite lateral sides were painted with a conductive carbon paint. Then, copper tape was pasted onto the first conductive layer in order to gain good contact of the samples with a power supply and wattmeter (self-heating experiment) and multimeter (electrical properties).

Resistance R (Ω) of the dried samples was measured by a Fluke 8846A $6\frac{1}{2}$ digit precise multimeter, and the electrical conductivity σ (S·m^{-1}) was calculated with respect to the shape ratio of the samples (electrodes: 50×50 mm^2, distance between electrodes: 50 mm) by the following equation

$$\sigma = \frac{1}{R} \cdot \frac{l}{S}, \qquad (2)$$

where R (Ω) is the resistance of the sample, l (m) is the distance between electrodes, and S (m^2) is the area of electrodes.

Self-heating experiments were performed on samples with dimensions of $50 \times 50 \times 50$ mm^3. Dried samples with electrodes attached in the same way as for determining the electrical properties were connected to a GW Instek GPR-11H30D voltage power supply and loaded by one or two voltage levels (Figure 2). Mortar samples CB 0 and CB 0.75 were loaded by 40 V to prove the no self-heating ability that was expected because of the electrical conductivity measurements. CB 1.25 mortar and mortars with higher amounts of CB were tested at two voltage levels, 40 and 100 V. Electrical power was monitored by a GW Instek GPM-8213 wattmeter. Ambient laboratory temperature and temperatures of samples were monitored by Pt-100 probes supported by Ahlborn ALMEMO 8690-9A datalogger in the central position of the bottom sides of the samples perpendicular to the attached electrodes.

(a)   (b)

Figure 2. (**a**) Samples for self-heating experiments. (**b**) Apparatus for self-heating experiments.

3. Results

3.1. Basic Physical Properties

In Table 2, the basic physical properties were represented by the bulk density, the matrix density, and the total open porosity. The bulk density was highest for the reference mortar CB 0 (2111 kg·m^{-3}) and decreased systematically with increasing CB amounts down to 1720 kg·m^{-3} (CB 2.25). The matrix density was in the range of 2562–2602 kg·m^{-3} and did not exhibit a significant influence on the amount of CB admixture. The total open porosity was calculated from the bulk density and the matrix density by Equation (1). It was lowest for CB 0 (17.6%) and exhibited an increasing tendency with increasing amounts of CB up to 33.4% (CB 2.25). Compared to the reference mortar, the addition of 0.75 wt. % of CB resulted in an increase in porosity of 7.6%, whereas addition of 2.25% wt. % of CB almost doubled the porosity (increase of 15.8%).

Table 2. Basic physical properties of the studied geopolymers.

	Bulk Density (kg·m^{-3})	Matrix Density (kg·m^{-3})	Total Open Porosity (%)
CB 0	2111	2562	17.6
CB 0.75	1947	2602	25.2
CB 1.25	1914	2588	26.0
CB 1.5	1816	2577	29.5
CB 1.75	1773	2568	31.0
CB 2	1749	2570	31.9
CB 2.25	1720	2582	33.4

3.2. Mechanical Properties

In Table 3, mechanical properties were represented by the compressive and the flexural strength. The highest compressive strength was observed for the reference mortar (83.45 MPa) and decreased systematically with an increasing amount of CB to approximately 7.3 MPa (CB 2 and CB 2.25). The compressive strength of CB 1.5 was about 55% that of the reference mortar, whereas CB 2 and CB 2.25 were about 9%. The highest flexural strength was observed for the reference mortar (8.26 MPa) and decreased systematically with an increasing amount of CB to 2.4 MPa (CB 2.25), which was about 30% of the value for the reference mortar.

Table 3. Mechanical properties of the studied geopolymers.

	Compressive Strength (MPa)	Flexural Strength (MPa)
CB 0	83.45	8.26
CB 0.75	68.60	5.23
CB 1.25	46.49	5.21
CB 1.5	20.38	5.04
CB 1.75	17.00	3.75
CB 2	7.31	3.85
CB 2.25	7.31	2.40

3.3. Thermal and Electrical Properties

In Table 4, thermal properties were represented by the thermal conductivity, and the specific heat capacity and electrical properties were represented by the electrical conductivity. The highest thermal conductivity was observed for CB 0 ($\lambda = 1.71$ W·m^{-1}·K^{-1}). The decreasing tendency of the thermal conductivity with an increasing amount of CB admixture corresponds with the basic physical properties (the bulk density was directly proportional, and the total open porosity was inversely proportional). CB 2.25 exhibited the lowest thermal conductivity (0.63 W·m^{-1}·K^{-1}). The specific heat capacity of the mortars was in the range of 715–849 J·kg^{-1}·K^{-1}. The maximum value (849 J·kg^{-1}·K^{-1}) was observed for CB 2.25. Higher values around 800 J·kg^{-1}·K^{-1} were observed for the reference mortar CB 0 and the mortars with higher dosages of CB (CB 1.75, CB 2), whereas lower values (715–733 J·kg^{-1}·K^{-1}) were observed for mortars with lower dosages of CB (CB 0.75, CB 1.25, CB 1.5).

Table 4. Thermal and electrical properties of the studied geopolymers.

	Thermal Conductivity (W·m^{-1}·K^{-1})	Specific Heat Capacity (J·kg^{-1}·K^{-1})	Electrical Conductivity (S·m^{-1})
CB 0	1.71	790	8.0×10^{-7}
CB 0.75	0.94	715	2.7×10^{-5}
CB 1.25	0.85	733	8.6×10^{-5}
CB 1.5	0.81	728	3.2×10^{-3}
CB 1.75	0.75	803	1.1×10^{-2}
CB 2	0.71	792	2.2×10^{-2}
CB 2.25	0.63	849	1.3×10^{-1}

The electrical conductivity increased significantly with an increasing amount of CB because of the formations of conductive paths within the geopolymer matrix. Such an increase is a very important assumption for self-heating ability of the tested mortars. The initial value of the electrically non-conductive mortar CB 0 (8×10^{-7} Ω·m) improved to the highest value (1.3×10^{-1} Ω·m) observed for CB 2.25, which was an increase of about 7 orders of magnitude. With respect to the measured data, it was expected that the percolation threshold for the self-heating ability would be at 1.5 wt. % of CB. The electrical conductivity of CB 1.5 mortar was 4000 times higher than that of the reference mortar (3.2×10^{-3} Ω·m).

3.4. Self-Heating Ability

In Figures 3–14, self-heating experiments are presented. Each self-heating experiment involved the determination of time dependencies of the ambient temperature, the temperature of the sample, and the power (dependent on power supply voltage and the passing current). Self-heating experiments conducted at 40 V on CB 0, CB 0.75, and CB 1.25 (Figures 3–5) and at 100 V on CB 1.25 (Figure 6) confirmed the expectations obtained by measurements of electrical properties that such materials did not exhibit a self-heating ability. The passing current, which is directly proportional to the measured power *P* (W), was negligible or very low in evolving the Joule heat. Heating was not possible even after the voltage increased from 40 to 100 V.

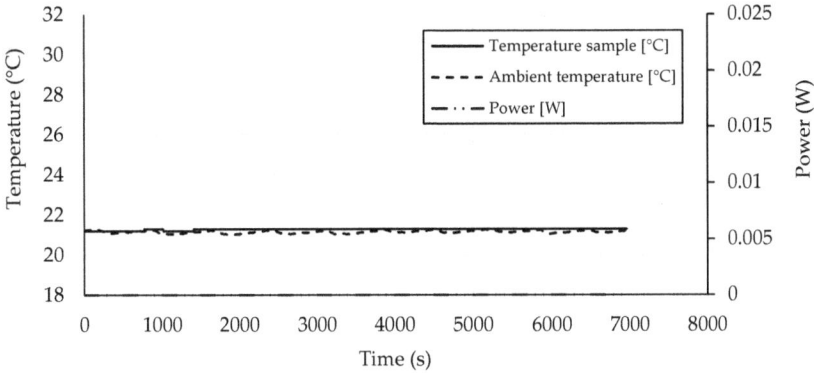

Figure 3. Self-heating experiment—CB 0, 40 V.

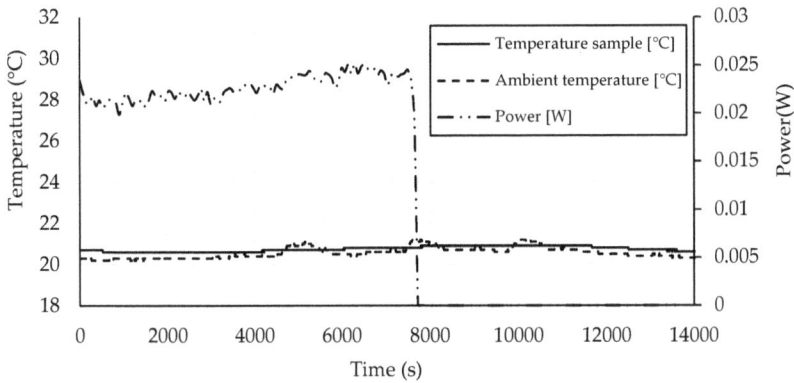

Figure 4. Self-heating experiment—CB 0.75, 40 V.

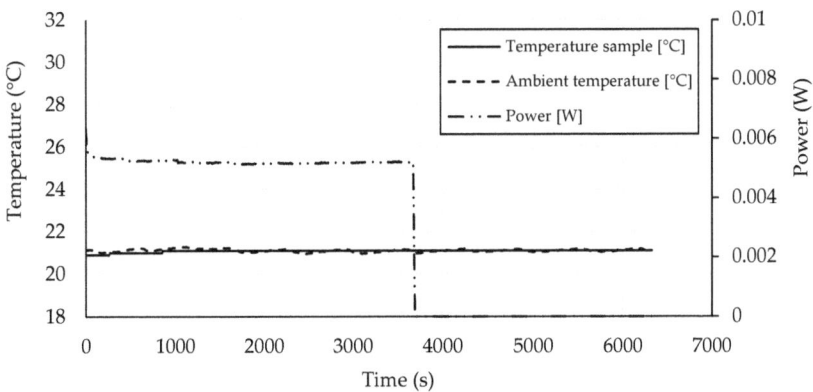

Figure 5. Self-heating experiment—CB 1.25, 40 V.

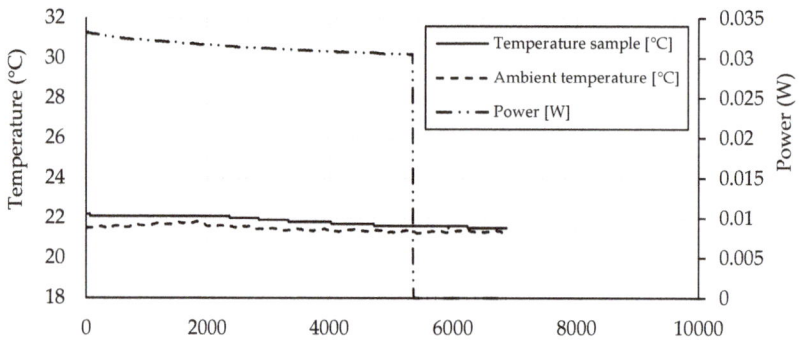

Figure 6. Self-heating experiment—CB 1.25, 100 V.

A low self-heating ability was observed for CB 1.5 at 40 V, where $\Delta t \approx 2$ °C and $P \approx 0.21$ W (Figure 7). An increased passing current induced by an increase of the voltage level from 40 to 100 V resulted in a temperature increase of $\Delta t \approx 9.5$ °C with the corresponding power $P \approx 1.18$ W (Figure 8).

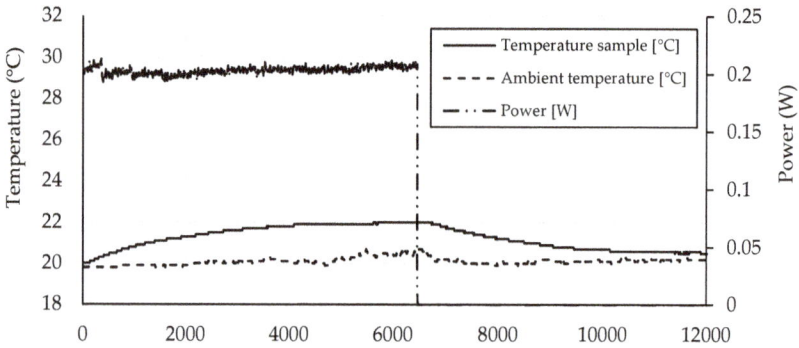

Figure 7. Self-heating experiment—CB 1.5, 40 V.

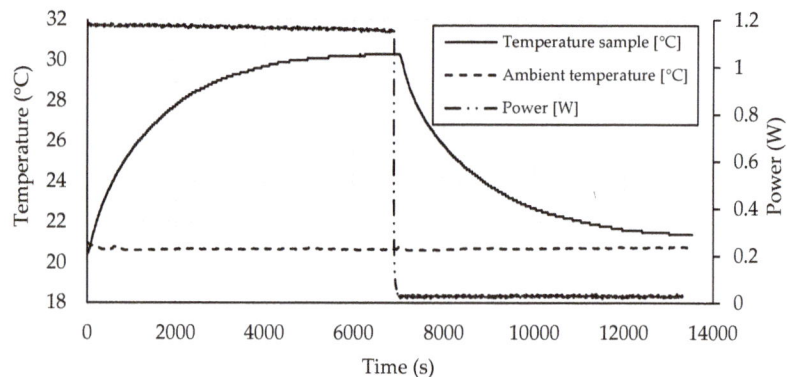

Figure 8. Self-heating experiment—CB 1.5, 100 V.

CB 1.75 exhibited a slightly better self-heating performance than the geopolymer mortars with a lower amount of CB. At 40 V, $\Delta t \approx 3$ °C and $P \approx 0.51$ W were achieved (Figure 9). At 100 V, further increases in the temperature $\Delta t \approx 22$ °C and the power $P \approx 3.45$ W were observed (Figure 10). CB 2 exhibited a good self-heating performance at 40 V, where $\Delta t \approx 10\ °C$ and $P \approx 1.25$ W (Figure 11), and at 100 V, where $\Delta t \approx 50\ °C$ and $P \approx 7.41$ W were achieved (Figure 12).

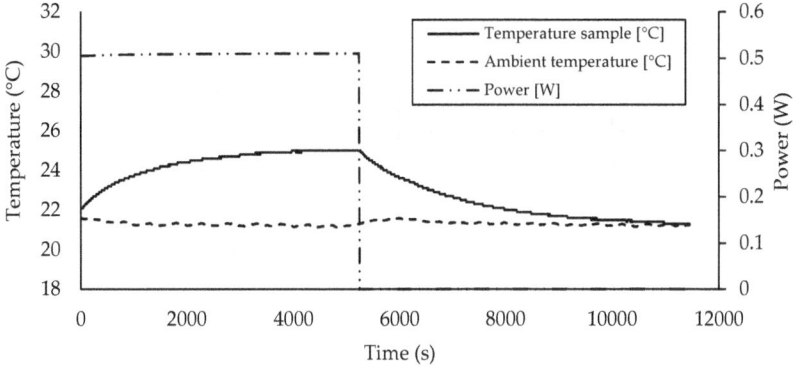

Figure 9. Self-heating experiment—CB 1.75, 40 V.

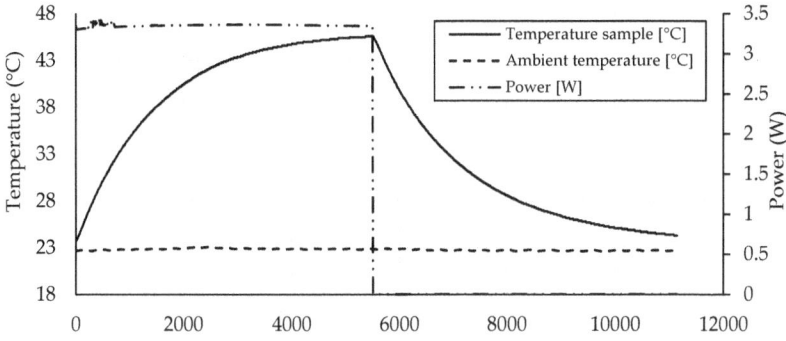

Figure 10. Self-heating experiment—CB 1.75, 100 V.

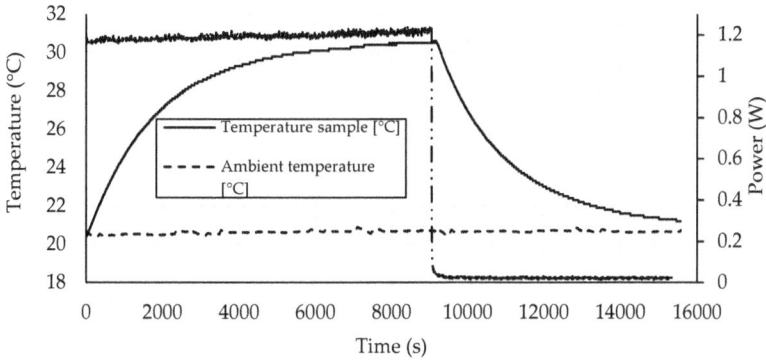

Figure 11. Self-heating experiment—CB 2, 40 V.

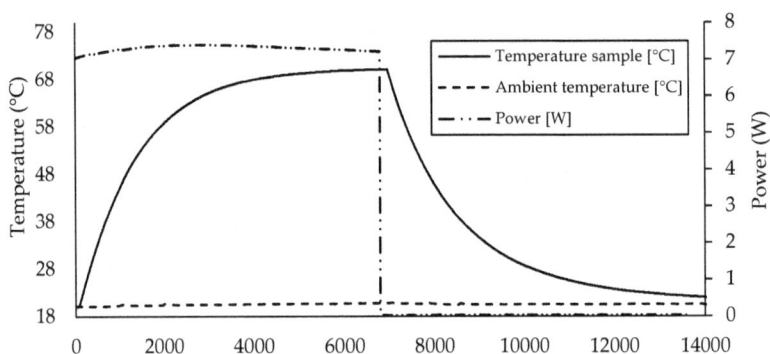

Figure 12. Self-heating experiment—CB 2, 100 V.

CB 2.25 exhibited the best self-heating performance. At 40 V, $\Delta t \approx 26\ °C$ and $P \approx 3.63$ W were achieved (Figure 13), and at 100 V the temperature increase was $\Delta t \approx 110\ °C$ and the corresponding power P was 25.99 W (Figure 14).

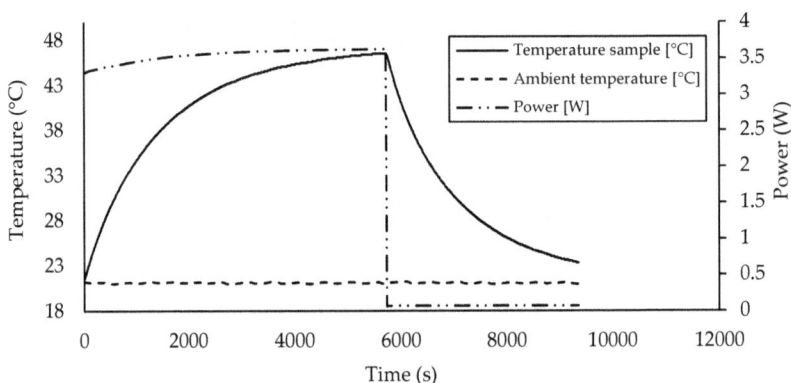

Figure 13. Self-heating experiment—CB 2.25, 40 V.

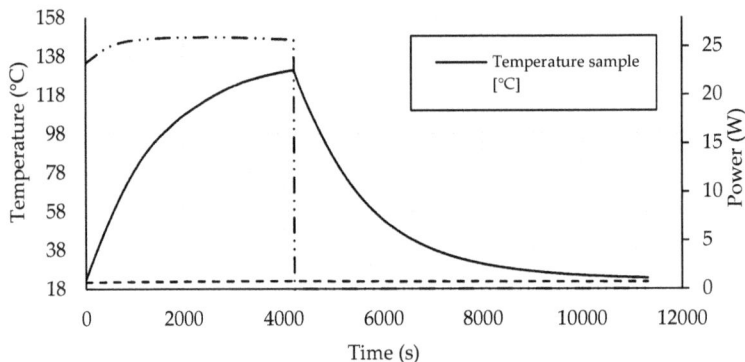

Figure 14. Self-heating experiment—CB 2.25, 100 V.

In Figure 15a, the maximal values of power from the conducted self-heating experiments loaded by 40 and 100 V are summarized. Geopolymer mortars with an amount of CB admixture starting at 1.5 wt. % were able to generate heat, leading to a temperature increase at both voltage levels. Comparing the heating power at 40 and 100 V of the mortars able to evolve heat, the highest increase was achieved for CB 2.25 (7.2 times higher power at 100 V than at 40 V), whereas the lowest non-zero increase was for CB 1.5 (5.6 times higher power at 100 V than at 40 V). In Figure 15b, the maximal achieved temperatures of the mortars during the self-heating experiments and the compressive strengths are presented. Quantities exhibited the opposite trend, where higher temperatures were achieved with an increasing amount of CB, but mechanical properties deteriorated.

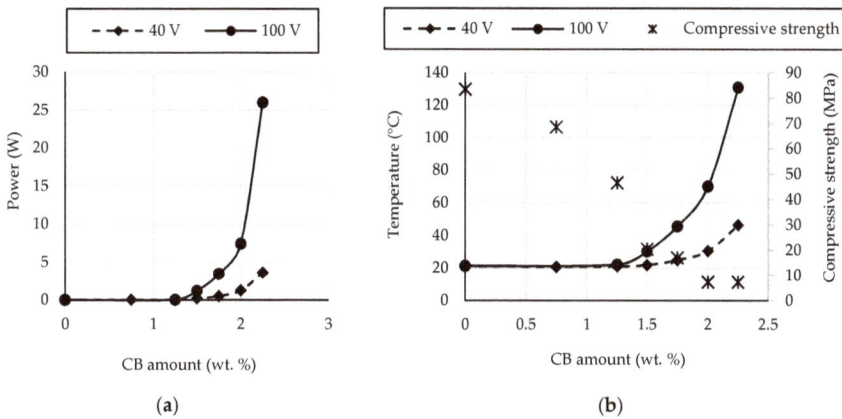

Figure 15. Dependence of (**a**) the power on CB amount (40 V, 100 V); (**b**) maximal achieved temperature and the compressive strength on CB amount (40 V, 100 V).

4. Discussion

With respect to the data presented in Tables 2 and 3, decreases in the bulk density, compressive strength, and flexural strength and an increase in the total open porosity of the designed geopolymer mortars were observed with an increasing amount of CB admixture. This mainly was due to the high surface area of particles and aggregated clusters of particles of such ECA filler (Figure 1). The average grain size of CB was $d_{50} = 0.52$ μm and $d_{90} = 17.6$ μm, which was significantly lower compared to the grain size of the slag binder ($d_{50} = 15.5$ μm and $d_{90} = 38.3$ μm). Because of the high surface area of CB, more mixing water was needed (Table 1), which resulted in an increase in the total open porosity. However, an increased amount of water was necessary for the preparation of mixtures with plastic consistencies with an average base diameter equal to 160 mm according to the ČSN EN 1015-3 standard Determination of Consistence of Fresh Mortar by Flow Table. The highest decrease in bulk density and the highest increase in total open porosity observed between the reference mortar (CB 0) and the mortar with the highest amount of CB (CB 2.25) were about 18.5% and 90%.

Mechanical properties of the mortars with higher CB dosages were negatively influenced by an increased total open porosity. The compressive strength of CB 0 (83.45 MPa) was significantly higher compared to that of the mortars with the self-heating ability. The decrease was significant (CB 1.5 about 76% compared to CB 0), but remained at a good level (20.38 MPa up to 1.5 wt. % of CB admixture). Geopolymer mortars with an amount of CB higher than 1.75 wt. % exhibited low compressive strength, equal to 7.31 MPa, which was about a 91% decrease compared to the reference mortar. However, it should be noted that the compressive strength of cementitious mortars widely used in practice with aggregates up to 2 mm is usually up to 10 MPa. In the case of the flexural strength, the highest value was observed for CB 0 (8.26 MPa) and decreased with an increasing amount of CB, but not as much as in the case of the compressive strength (CB 2.25 compared to CB 0, about a 71% decrease). Taking into consideration cement mortars widely used in practice with flexural strengths up to 2.5 MPa, all the designed geopolymer mortars are comparable.

The thermal conductivity, an important parameter describing the ability of effective spreading of the generated Joule heat, exhibited a decreasing tendency with an increasing amount of CB, which was influenced, again, by the increasing porosity. The highest thermal conductivity was observed for CB 0 (1.71 $W \cdot m^{-1} \cdot K^{-1}$), whereas the lowest was observed for CB 2.25 (0.63 $W \cdot m^{-1} \cdot K^{-1}$). The thermal conductivity of all the mortars with CB admixtures was below 1 $W \cdot m^{-1} \cdot K^{-1}$. The specific heat capacity of the mortars ranged between 715–849 $J \cdot kg^{-1} \cdot K^{-1}$. Higher values around 800 $J \cdot kg^{-1} \cdot K^{-1}$ were observed for the reference mortar CB 0 and the mortars with higher dosages of CB (CB 1.75, CB 2), whereas lower values (715– 733 $J \cdot kg^{-1} \cdot K^{-1}$) were observed for mortars with lower dosages of CB (CB 0.75, CB 1.25, CB 1.5). The maximum value (849 $J \cdot kg^{-1} \cdot K^{-1}$) was observed for CB 2.25.

Electrical properties represented by the electrical conductivity were essential in terms of the main aim of this paper, which was the design of multifunctional geopolymers. The self-heating ability can be achieved just by significantly increasing the electrical conductivity. The reference mortar was a typical electrical insulator with an electrical conductivity of 8.0×10^{-7} $S \cdot m^{-1}$; therefore, it was not able to generate Joule heat. The electrical conductivity increased by 1 order of magnitude for CB 0.75 and 2 orders of magnitude for CB 1.25, which was not sufficient. The electrical conductivity of CB 1.5 increased by about 3 orders of magnitude, which was close to the percolation threshold, and the slight self-heating ability of this mortar was further observed. Further improvement was observed for CB 1.75 (4 orders of magnitude), CB 2 (4 orders of magnitude), and CB 2.25 (5 orders of magnitude), which demonstrated a significant enhancement of electrical properties.

Self-heating experiments proved a slight self-heating ability of CB 1.5 at 40 V, and a power 0.21 W was able to heat up the sample by about 2 °C. At 40 V loading, the power increased with increasing the amount of CB in the following way: CB 1.75, 0.51 W; CB 2, 1.25 W; and CB 2.25, 3.63 W; with corresponding temperature increases of CB 1.75, 3 °C; CB 2, 10 °C; and CB 2.25, 26 °C. At 100 V loading, the powers of CB 1.5, CB 1.75, CB 2, and CB 2.25 were 1.18, 3.45, 7.41, and 25.99 W, and temperature increases were 9.5, 22, 50, and 110 °C, respectively. It is evident that an increase in the applied voltage from 40 to 100 V leads to a significantly higher self-heating ability. In Figure 15b, the maximal achieved temperatures are presented together with the compressive strength dependent on the amount of CB. Despite the fact that the mechanical properties of geopolymer mortars with the CB admixture are significantly lower compared to the reference mortar, their mechanical properties are comparable to, or even better than, commonly used cementitious mortars, and such types of material can find applications in the construction industry.

5. Conclusions

A conducted investigation on the basic physical, mechanical, thermal, electrical properties, and the self-heating ability of alkali-activated slag mortars with carbon black as a conductive filler is presented in this paper, and the following conclusions have been drawn from the experimental results:

- An increase in the amount of CB admixture in geopolymers based on GBFS activated by water glass led to the deterioration of mechanical properties, which was attributed to an increased amount of mixing water and, consequently, increased porosity.
- An increase in the amount of CB admixture in geopolymers based on GBFS activated by water glass led to a decrease in thermal conductivity, which is an important parameter describing the ability to spread the evolved heat.
- The percolation threshold for the self-heating ability was around 1.5 wt. % of CB, where a slight self-heating ability was observed.
- Geopolymers based on GBFS activated by water glass with CB amounts in the range of 1.75–2.25 wt. % exhibited good self-heating abilities.
- The self-heating ability of geopolymers based on GBFS activated by water glass with CB can be significantly improved by increasing the voltage. The heating power of the geopolymer mortar

with CB in the amount of 2.25 wt. % at 40 V was similar to the heating power of the geopolymer mortar with CB in the amount of 1.75 wt. % at 100 V (3.63 vs. 3.45 W).

This study proved the possibility to design multifunctional geopolymers with self-heating abilities based on alkali-activated GBFS and CB admixture. However, further investigation is necessary, especially in terms of optimizing the designed mixtures leading to a decrease in the porosity and in an effective homogenization of CB, which will ensure maximization of the self-heating ability with a lower deterioration of mechanical properties.

Author Contributions: Writing—original draft, Methodology, Resources, L.F.; Experimental—electrical properties, thermal properties, self-heating experiment, M.P.; Methodology, Resources, W.-T.L.; Experimental—basic physical properties, mechanical properties, L.P.; Supervision, R.Č.

Funding: This research was funded by the Czech Science Foundation under the project No. 19-11516S, Ministry of Science and Technology (MOST, Taiwan) under the project MOST-108-2221-E-197-006, and Ministry of Industry and Trade of the Czech Republic under innovation voucher No. CZ.011.02/0.0/17_115/0012287.

Conflicts of Interest: The authors declare no conflicts of interest.

References

1. Han, B.G.; Wang, Y.Y.; Dong, S.F.; Zhang, L.Q.; Ding, S.Q.; Yu, X.; Ou, J.P. Smart concretes and structures: A review. *J. Intell. Mater. Syst. Struct.* **2015**, *26*, 1303–1345. [CrossRef]
2. Rana, S.; Subramani, P.; Fangueiro, R.; Correia, A.G. A review on smart self-sensing composite materials for civil engineering applications. *Aims Mater. Sci.* **2016**, *3*, 357–379. [CrossRef]
3. Han, B.; Sun, S.; Ding, S.; Zhang, L.; Yu, X.; Ou, J. Review of nanocarbon-engineered multifunctional cementitious composites. *Compos. Part A Appl. Sci. Manuf.* **2015**, *70*, 69–81. [CrossRef]
4. Li, Z.; Ding, S.; Yu, X.; Han, B.; Ou, J. Multifunctional cementitious composites modified with nano titanium dioxide: A review. *Compos. Part A: Appl. Sci. Manuf.* **2018**, *111*, 115–137. [CrossRef]
5. Pisello, A.L.; D'Alessandro, A.; Sambuco, S.; Rallini, M.; Ubertini, F.; Asdrubali, F.; Materazzi, A.L.; Cotana, F. Multipurpose experimental characterization of smart nanocomposite cement-based materials for thermal-energy efficiency and strain-sensing capability. *Sol. Energy Mater. Sol. Cells* **2017**, *161*, 77–88. [CrossRef]
6. Chung, D.D.L. Self-heating structural materials. *Smart Mater. Struct.* **2004**, *13*, 562–565. [CrossRef]
7. Gomis, J.; Galao, O.; Gomis, V.; Zornoza, E.; Garces, P. Self-heating and deicing conductive cement. Experimental study and modeling. *Constr. Build. Mater.* **2015**, *75*, 442–449. [CrossRef]
8. Armoosh, S.R.; Oltulu, M. Self-heating of electrically conductive metal-cementitious composites. *J. Intell. Mater. Syst. Struct.* **2019**, *30*, 2234–2240. [CrossRef]
9. Armoosh, S.R.; Oltulu, M. Effect of Different Micro Metal Powders on the Electrical Resistivity of Cementitious Composites. In Proceedings of the 3rd World Multidisciplinary Civil Engineering, Architecture, Urban Planning Symposium, Prague, Czech Republic, 18–22 June 2018; Volume 471.
10. Wei, J.; Zhao, L.; Zhang, Q.; Nie, Z.; Hao, L. Enhanced thermoelectric properties of cement-based composites with expanded graphite for climate adaptation and large-scale energy harvesting. *Energy Build.* **2018**, *159*, 66–74. [CrossRef]
11. Wei, J.; Fan, Y.; Zhao, L.; Xue, F.; Hao, L.; Zhang, Q. Thermoelectric properties of carbon nanotube reinforced cement-based composites fabricated by compression shear. *Ceram. Int.* **2018**, *44*, 5829–5833. [CrossRef]
12. Wei, J.; Nie, Z.; He, G.; Hao, L.; Zhao, L.; Zhang, Q. Energy harvesting from solar irradiation in cities using the thermoelectric behavior of carbon fiber reinforced cement composites. *RSC Adv.* **2014**, *4*, 48128–48134. [CrossRef]
13. Ali, M.; Saidur, R.; Hossain, M. A review on emission analysis in cement industries. *Renew. Sustain. Energy Rev.* **2011**, *15*, 2252–2261. [CrossRef]
14. Wei, J.X.; Cen, K.; Geng, Y.B. Evaluation and mitigation of cement CO_2 emissions: Projection of emission scenarios toward 2030 in China and proposal of the roadmap to a low-carbon world by 2050. *Mitig. Adapt. Strateg. Glob. Chang.* **2019**, *24*, 301–328. [CrossRef]
15. Andrew, R.M. Global CO_2 emissions from cement production, 1928–2017. *Earth Syst. Sci. Data* **2018**, *10*, 2213–2239. [CrossRef]

16. Mehta, A.; Siddique, R. An overview of geopolymers derived from industrial by-products. *Constr. Build. Mater.* **2016**, *127*, 183–198. [CrossRef]

17. Ma, C.-K.; Awang, A.Z.; Omar, W. Structural and material performance of geopolymer concrete: A review. *Constr. Build. Mater.* **2018**, *186*, 90–102. [CrossRef]

18. Albitar, M.; Ali, M.M.; Visintin, P.; Drechsler, M. Durability evaluation of geopolymer and conventional concretes. *Constr. Build. Mater.* **2017**, *136*, 374–385. [CrossRef]

19. Zuda, L.; Drchalova, J.; Rovnanik, P.; Bayer, P.; Keršner, Z.; Cerny, R. Alkali-activated aluminosilicate composite with heat-resistant lightweight aggregates exposed to high temperatures: Mechanical and water transport properties. *Cem. Concr. Compos.* **2010**, *32*, 157–163. [CrossRef]

20. Nguyen, T.H.Y.; Tsuchiya, K.; Atarashi, D. Microstructure and composition of fly ash and ground granulated blast furnace slag cement pastes in 42-month cured samples. *Constr. Build. Mater.* **2018**, *191*, 114–124. [CrossRef]

21. Davidovits, J. Geopolymers: Ceramic-Like Inorganic Polymers. *J. Ceram. Sci. Technol.* **2017**, *8*, 335–350.

22. Pacheco-Torgal, F.; Castro-Gomes, J.; Jalali, S. Alkali-activated binders: A review Part 1. Historical background, terminology, reaction mechanisms and hydration products. *Constr. Build. Mater.* **2008**, *22*, 1305–1314. [CrossRef]

23. Altan, E.; Erdoğan, S.T. Alkali activation of a slag at ambient and elevated temperatures. *Cem. Concr. Compos.* **2012**, *34*, 131–139. [CrossRef]

24. Ben Haha, M.; Le Saoût, G.; Winnefeld, F.; Lothenbach, B. Influence of activator type on hydration kinetics, hydrate assemblage and microstructural development of alkali activated blast-furnace slags. *Cem. Concr. Res.* **2011**, *41*, 301–310. [CrossRef]

25. Pacheco-Torgal, F.; Castro-Gomes, J.; Jalali, S. Alkali-activated binders: A review Part 2. About materials and binders manufacture. *Constr. Build. Mater.* **2018**, *22*, 1315–1322. [CrossRef]

26. Tang, Z.; Li, W.; Hu, Y.; Zhou, J.L.; Tam, V.W. Review on designs and properties of multifunctional alkali-activated materials (AAMs). *Constr. Build. Mater.* **2019**, *200*, 474–489. [CrossRef]

27. Rovnaník, P.; Kusák, I.; Bayer, P.; Schmid, P.; Fiala, L. Comparison of electrical and self-sensing properties of Portland cement and alkali-activated slag mortars. *Cem. Concr. Res.* **2019**, *118*, 84–91. [CrossRef]

28. Rovnanik, P.; Kusak, I.; Bayer, P.; Schmid, P.; Fiala, L. Electrical and Self-Sensing Properties of Alkali-Activated Slag Composite with Graphite Filler. *Materials* **2019**, *12*, 1616. [CrossRef]

29. Fiala, L.; Rovnanik, P.; Cerny, R. Investigation of the Joule's effect in electrically enhanced alkali-activated aluminosilicates. *Cem. Wapno Beton* **2017**, *22*, 201–210.

30. Dhoble, Y.N.; Ahmed, S. Review on the innovative uses of steel slag for waste minimization. *J. Mater. Cycles Waste Manag.* **2018**, *20*, 1373–1382. [CrossRef]

31. Guo, J.; Bao, Y.; Wang, M. Steel slag in China: Treatment, recycling, and management. *Waste Manag.* **2018**, *78*, 318–330. [CrossRef]

32. Cabot Company. Cabot launches new carbon black products for tyre applications. *Addit. Polym.* **2014**, *2014*, 2. [CrossRef]

33. Li, Y.C.; Huang, X.R.; Zeng, L.J.; Li, R.F.; Tian, H.F.; Fu, X.W.; Wang, Y.; Zhong, W.H. A review of the electrical and mechanical properties of carbon nanofiller-reinforced polymer composites. *J. Mater. Sci.* **2019**, *54*, 1036–1076. [CrossRef]

34. Monteiro, A.O.; Cachim, P.B.; Costa, P.M. Self-sensing piezoresistive cement composite loaded with carbon black particles. *Cem. Concr. Compos.* **2017**, *81*, 59–65. [CrossRef]

35. Fiala, L.; Jerman, M.; Rovnaník, P.; Černý, R. Basic physical, mechanical and electrical properties of electrically enhanced alkali-activated aluminosilicates. *Mater. Tehnol.* **2017**, *51*, 1005–1009. [CrossRef]

36. Alonso, M.; Gismera, S.; Blanco, M.; Lanzón, M.; Puertas, F. Alkali-activated mortars: Workability and rheological behaviour. *Constr. Build. Mater.* **2017**, *145*, 576–587. [CrossRef]

MDPI

St. Alban-Anlage 66

4052 Basel

Switzerland

Tel. +41 61 683 77 34

Fax +41 61 302 89 18

www.mdpi.com

Energies Editorial Office

E-mail: energies@mdpi.com

www.mdpi.com/journal/energies

www.ingramcontent.com/pod-product-compliance
Lightning Source LLC
Chambersburg PA
CBHW051913210326
41597CB00033B/6128